AF763568

Homely Atmospheres and Lighting Technologies in Denmark

HOME

Series Editors: Rosie Cox and Victor Buchli

ISSN: 2398–3191

This exciting new series responds to the growing interest in the home as an area of research and teaching. Highly interdisciplinary, titles feature contributions from across the social sciences, including anthropology, material culture studies, architecture and design, sociology, gender studies, migration studies, and environmental studies. Relevant to undergraduate and postgraduate students as well as researchers, the series will consolidate the home as a field of study.

Food, Masculinities, and Home: Interdisciplinary Perspectives, edited by Michelle Szabo and Shelley Koch
Living with Strangers: Bedsits and Boarding Houses in Modern English Life, Literature and Film, edited by Chiara Briganti and Kathy Mezei
Making Homes, Sarah Pink, Kerstin Leder Mackley, Roxana Morosanu, Val Mitchell, and Tracy Bhamra
Sexuality and Gender at Home: Experience, Politics, Transgression, edited by Brent Pilkey, Rachael M. Scicluna, Ben Campkin, and Barbara Penner
Thinking Home: Interdisciplinary Dialogues, edited by Sanja Bahun and Bojana Petric
Queering the Interior, edited by Andrew Gorman-Murray and Matt Cook

FURTHER TITLES FORTHCOMING

A Cultural History of Twin Beds, Hilary Hinds
Art and Masculinity in Post-War Britain, Greg Salter
An Anthropology of Home and Mobility in Europe, Sara Bonfanti, Aurora Massa, Paolo Boccagni and Alejandro Miranda

Homely Atmospheres and Lighting Technologies in Denmark

Living with Light

MIKKEL BILLE

BLOOMSBURY ACADEMIC
LONDON • NEW YORK • OXFORD • NEW DELHI • SYDNEY

BLOOMSBURY ACADEMIC
Bloomsbury Publishing Plc
50 Bedford Square, London, WC1B 3DP, UK
1385 Broadway, New York, NY 10018, USA

BLOOMSBURY, BLOOMSBURY ACADEMIC and the Diana logo are trademarks of Bloomsbury Publishing Plc

First published in Great Britain, 2019

Copyright © Mikkel Bille, 2019

Mikkel Bille has asserted his right under the Copyright, Designs and Patents Act, 1988, to be identified as Author of this work.

For legal purposes the Acknowledgements on p. vii constitute an extension of this copyright page.

Series design by Clare Turner

Cover image: © Federica Giusti on Unsplash

All rights reserved. No part of this publication may be reproduced or transmitted in any form or by any means, electronic or mechanical, including photocopying, recording, or any information storage or retrieval system, without prior permission in writing from the publishers.

Bloomsbury Publishing Plc does not have any control over, or responsibility for, any third-party websites referred to or in this book. All internet addresses given in this book were correct at the time of going to press. The author and publisher regret any inconvenience caused if addresses have changed or sites have ceased to exist, but can accept no responsibility for any such changes.

A catalogue record for this book is available from the British Library.

A catalog record for this book is available from the Library of Congress.

ISBN: HB: 978-1-3500-5718-0
ePDF: 978-1-3500-5719-7
ePub: 978-1-3500-5720-3

Series: Home

Typeset by Refine Catch Limited, Bungay, Suffolk
Printed and bound in Great Britain

To find out more about our authors and books visit www.bloomsbury.com and sign up for our newsletters

Contents

Figures

Acknowledgements

Many people have inspired and shaped my thinking about light over the years. Among these are Victor Buchli, who encouraged my early writings on illumination, and the collaboration with Tim Flohr Sørensen, which fundamentally shaped my approach to luminosity. Also acknowledged are the many people who have commented on the manuscript or chapters, and discussed ideas in this book; in particular I would like to thank Andreas Bandak, Bettina Hauge, Casper Ebbensgaard, Tim Edensor, Jonas Larsen, Sarah Pink, Siri Schwabe, Tim Flohr Sørensen, and Pernille Wiil.

The main part of this book was written during a three-month visiting fellowship at the Demand Centre at Lancaster University, where I thank Elizabeth Shove and Gordon Walker for providing such a memorable stay. Here Mette Kragh-Furbo, Stanley Blue, Torik Holmes, Noel Cass, Janine Morley, Yolande Strengers, and others affiliated gave constructive comments on my work in progress. Also, discussions with Don Slater during a visiting fellowship at LSE have been much appreciated.

Selected parts of Chapters 3, 4 and 5 have been published in different forms in Bille, M. (2015). Hazy worlds: Atmospheric ontologies in Denmark. *Anthropological Theory, 15*(3), 257–274; and Bille, M. (2015). Lighting up cosy atmospheres in Denmark. *Emotion, Space and Society, 15*, 56–63. Paragraphs from Chapter 8 have been published in Bille, M. (2013). Luminous atmospheres: Energy politics, climate technologies, and cosiness in Denmark. *Ambiances. International Journal of Sensory Environment, Architecture and Urban Space.*

Funding for my research has been provided by the Danish Council for Independent Research (*Everyday life of Light*, 2010–2013), the Carlsberg Foundation (*Light Culture* conference in 2015), and the Velux foundation in the final stages (*Living with Nordic lighting*, 2018–2021).

Series preface: why home?

Rosie Cox and Victor Buchli

The home is where people are made and undone. As life is increasingly seen as precarious, fluid, mobile and globalized, there is a growing interest in the home: what it is, what it means to various groups of people, how it constitutes them and how it relates to other spheres of life both in the present and in the past. Home is both physical and metaphorical, local and national, a place of belonging and exclusion. It is at the heart of the most seemingly mundane spaces and experiences – the site of quotidian activities such as eating, washing, raising children and loving. Yet it is precisely the purportedly banal nature of the home that masks its deep importance for the underlying assumptions that structure social and political life. Home reveals the importance of routine activities, such as consumption, to highly significant and urgent wide-ranging issues and processes such as the maintenance of and challenges to global capitalism and our relationship to the natural environment.

Among academic writers home is increasingly problematized, interrogated and reconsidered. Long understood as an axis of gender inequality, home is also seen as a site; a space of negotiation and resistance as well as oppression, and a place where such relationships are undone as well as made. As a topic of study, it is the natural analytical unit for a number of disciplines, with relevance to a wide range of cultural and historical settings. The home is probably one of the few truly universal categories upon which an interdisciplinary programme of research can be conducted and which over recent years has resulted in a distinctive analytical category across disciplines, times and cultures.

This book series offers a space to foster these debates and to move forward our thinking about the home. The books range across the social and historical sciences, drawing out the cross-cutting themes and interrelationships within writings on home and providing us with new perspectives on this intimate space. While our understanding of 'home' is expansive, and open to interrogation, it is not unbounded. In honing our understandings of what 'home' is, this series aims to disturb and it goes beyond the domestic including sites and states of dispossession and homelessness and experiences of the 'unhomely'.

1

Introduction

Living with light

It is an autumn evening in Denmark, the sun is setting, and Cecilia is tidying her place. She makes sure there is no clutter, and that refreshments and sweets are ready for her friends, who are about to arrive. It is fairly common in Denmark to have friends visiting one's home, compared to countries with warmer weather and a more vibrant restaurant, pub or café culture. Many people in Denmark make great effort to orchestrate a homely atmosphere that is 'just right' for a successful social gathering. Candy, cake, candlelight, music, alcohol, tea and coffee are material props that help to make people feel at ease; at least this is the idea. Cecilia, like other Danes, puts a lot of effort into arranging her home with design objects, and, if space allows, she creates different spaces within rooms for different situations: a reading spot; a relaxing sofa arrangement; a dinner section, etc. With such emphasis on atmosphere, materiality and presentation of self, the home in Denmark is an important place for social interaction and forging identity (Højer & Vacher 2009; Philipsen 2013; Winther 2005, 2006).

In Cecilia's home, two unusually luminous ceiling lamps are lit (one seen in Figure 1.1). Reaching for the light switch on the wall is one of her most common, unconscious, bodily acts, as is the switching on of the pendant lamp above the dinner table, located just next to the kitchen, which spreads a dimmed, glare-free light. This is where she assumes – and wishes – the guests will sit on arrival. She similarly turns on the two pendant lamps above the sofa arrangement at the other end of the room, even though they will not sit there until later in the evening. The light in the reading spot is, however, not turned on. She stages the lighting in this way to create an atmosphere she calls *hygge*, commonly translated as cosiness, and a term describing an atmosphere seen as particularly 'Danish' (Bille & Sørensen 2007; Hansen 1976; Linnet 2011). Tea lights on the table remain unlit; she is afraid she will forget them and they will cause a fire. Yet they are still there, as a central element in the creation of

FIGURE 1.1 *Lighting above the dinner table and sofa in Cecilia's home. Photo by the author.*

a *hygge* atmosphere. As the evening progresses, her guests move from the dinner table part of the kitchen-cum-living room (measuring roughly 25m^2) to the sofa area for even more relaxed conversation. The last thing Cecilia will do before joining her guests on the sofa is to turn off the luminous ceiling lamps, while keeping the light over the dinner table at the distant end of the room on, in order to shape an even more intimate atmosphere through cosy lighting, or *hyggelys* as it is called. She would normally not turn the two ceiling lamps off if she was alone, but she presumes the guests would be annoyed by them while sitting on the sofa, and the room feels cosier (*hyggeligt*) with them turned off. After the guests have left, she cleans up the place, turns off the lights in the kitchen-cum-living room, and heads to bed. Naturally, light alone does not create *hygge*, but it is an important part of it, she explains. Light is an aspect of a homely atmosphere – an atmospheric technology.

Cecilia is actually a little dissatisfied with the light from the two pendant lamps above the sofa. Not the lamp design as such, which she likes, but the light they produce; it is a minor irritation stemming from the fact that incandescent light bulbs were banned in 2012 in the European Union and replaced by energy-saving light bulbs. She feels these new bulbs are not as 'good' as the incandescent ones she used to have because they have a different shape:

> the lamps hanging over the table are not made for those new [energy saving] round-topped bulbs; they were made for the [incandescent] flat-topped reflector bulbs. But those are not available anymore, and one has to be 'a good girl' and put the round-topped energy saving ones in instead. As you can tell, I am not satisfied at all. It's all well and good with green energy and the environment and all that, if only they could keep their hands off my incandescent light bulbs.

One would perhaps not assume it at first sight, but a conflict is developing here in something as common as lighting one's home. Staging the visual appearance of an evening in Cecilia's home on the one hand exemplifies the common practice of orchestrating homely atmospheres and cosy gatherings through lighting. But it also shows the dissatisfaction with the minor visual differences energy-saving light bulbs have enforced.

While lighting may be about technical quality of light, measurements and energy efficiency, conversations with people easily turn to half an hour's rumination on atmospheres, uncomfortable lighting in foreign countries, scary urban places, the bleak quality of the new energy-saving lighting, wrong bulb purchases and cosy (particle polluting) candle lights, which are called 'living lights' in Denmark. People may not have thought much about it until they are confronted by a different quality of illumination, or by an insistent anthropologist, but light is pivotal when shaping the atmosphere of a place. Informants are struck by the awkward sense of familiarity with what has been previously unnoticed, when I point out that in order to increase the flow of people on the Metro in Copenhagen bright light is used to help encourage movement along certain routes; or that there is a 'colder light' in the fish counter in the supermarket in order to make the fish look 'fresher'; or that pinkish light is used to stop teenagers congregating in certain public places in England because it makes acne more apparent – similar to the well-known blue lighting in public toilets that conceals the veins of drug users. Such examples show that lighting technologies help constitute social life by conducting politics through (intangible) material means, increase consumption, and guide behaviour, most often at the margins of people's attention. While people may more or less consciously orchestrate domestic lighting to fit activities and moods, public spaces are orchestrated visually for us at the edges of awareness to meet such political, economic or behavioural goals.

Light is so natural it is easy to forget how socially mediated it always is. A central analytical point in this book is that lighting is actually rarely about visibility. It may not even be so much about seeing in a strictly physiological sense. It is about *seeing and sensing in a particular way*. Lighting is an atmospheric element that tinges the material infrastructure and its affective presence, and thus also becomes appreciated in particular ways depending on context. Of course, when going up to a pitch-dark attic, lighting is about visibility, but how much light, the tolerance for glare and colour, and even what technology is available, is an outcome of social processes. Light is in essence thoroughly embedded in socio-material life, as much as it shapes the human body's instrumental ability to see the surrounding world. As Marcel Mauss (1973) notably commented, one never simply looks, moves or senses. Rather, one learns how to through practices in specific contexts. Light and lighting, as something so commonly experienced and recognized, are imbued in cultural ideas and norms about intimacy, caring for each other, social gatherings, political power, class, sense of security, and traditions of what counts as visual comfort. And such sensory norms and connotations are learned rather than given.

To illustrate this point, let us return to Cecilia: *Cecilia is visually impaired.* She sheds light on the world to shape a visual comfort she has never herself fully experienced, employing sighted people's vocabulary. She is not completely blind, though. While she has no sight, she does have some sense of light. She could find her way around the apartment in total darkness, but she got fed up of bumping into things, and found that sharp light offers a rough sense of orientation (hence the two luminous ceiling lamps were specifically selected for their light quality). At night, she also has a light turned on in the hallway between her bedroom and the toilet, so she can orient herself towards the door when waking up. But not just any sharp light will do. In some places, she could sense the difference in lighting quality between an incandescent and an energy-saving light bulb from the higher colour temperature; in other places, it was the spread of light the round-topped energy-saving light bulb gave, compared to the previous flat-topped incandescent reflector bulb where the lampshade orchestrated the light. As she said, 'I cannot see it on the lamp, but I can see it on the light, it falls in a different way, it spreads a bit more.'

While it is commonly noted that people do not *see* light, but *see in* light (Ingold 2000: 265, 2011: 134; Merleau-Ponty 1964: 178), perhaps it is with Cecilia fruitful to rethink that famed dictum. As Gernot Böhme (2017: chapter 20) has rightly asked, is it not light we see when we are blinded by the light? Is it not light we see, when it falls through the windows, reflecting on surfaces of dust particles or smoke. We may of course *know* that it is actually dust particles that the light beams touch and we see; but cognitively knowing and sensing are not the same. Cecilia may of course also *know* that the light source creates a physiological response in her visually impaired eyes, and this is what she 'sees' rather than light 'itself'. But perhaps, when we analyse how people sense the world around us – at least phenomenologically speaking – it is not fruitful to maintain such sharply delineated difference between objects, light source and reflecting objects, but instead approach a more vague, amalgamated, human embeddedness in illumination that rests on a long range of cultural logics and assumptions.

Cecilia's domestic illumination illustrates something fascinating about the cultural logic informing the way people use light at home, not just in order to see something, but to see it in 'the right way', to make a place 'feel right' (cf. Love 2016; Pink et al. 2015). 'Vision' as such a continuous and mundane practice is performed *with* light and not just *in* light. Yet while a phenomenological account may highlight this embeddedness – rejecting an *a priori* distinction between subject and object – it often ignores how such sensuous encounter is culturally informed and trained. As Cristina Grasseni notes, 'Vision, like the other senses, needs educating and training in a relationship of apprenticeship and within an ecology of practice' (2004: 41). Small changes in lighting technology can in this sense mean a lot to people's perception of visual comfort and norms. This book testifies to how the home is one site where such luminous training, perception and ideals take shape. Despite economic and ethical incentives to switch to energy-saving light bulbs, the background for this book is that many Danes have been reluctant to do so, hoarding incandescent bulbs, buying the less energy-efficient halogen bulbs, or using the new bulbs but complaining about poorer visual comfort. By being a location where people arrange themselves to feel at ease and well, the home here represents a 'lumitopia' (Edensor & Bille 2017) – a space in which an intensified attention to illumination is integral to the very particularity of the place in ways that makes sense – at least to them. This, of course, neither entails that people are always successful in orchestrating such spaces, or that the home is not also a stage for conflict and forging of moral subjects.

Time is working against the ethnographic descriptions presented here. Energy-saving light bulbs are constantly improving in quality, becoming more akin to incandescent lighting. Furthermore, many parts of the lighting industry are adopting the view that 'good lighting' is universal, circadian, and mimics the sun, and thus increasingly install dynamic lighting in offices and public institutions to produce better indoor environments. But what does that tell us about good *domestic* lighting and the sensory qualities related to it? It may be that light creates life, in contrast to darkness. But there is a wide spectrum between darkness and light, with natural and artificial light, and it is in the nuances of this spectrum, in the shadows, luminance, reflections and glare that people live. In Denmark, as shown with Cecilia, while light is life, it is in the moulded, dimmed light that people tend to live their lives. Despite rapid technological changes, light is embedded in atmospheric practices and cultural values, and these are only scarcely understood. This book is about such cultural notions and lighting practices. It illustrates the meaning light makes, the knowledge it takes, the politics it creates, and the concepts and practices lighting is shaped by, and conversely shapes.

Aims and scope

A large amount of research has studied how the physiological body and individual psychology reacts to various light, and its physical, spatial and environmental properties, mostly through positivist experiments and measurements (cf. Aries & Newsham 2008; Boyce 2014; Corrodi & Spechtenhauser 2008; Hopkinson 1964; Mills & Borg 1999). Such studies offer valuable insights that have shaped the way light is used and understood today, no doubt. But they are not sufficiently equipped to answer questions about the social life light partakes in shaping: the cultural logics and social practices embedded in the light shed on domestic and public spaces.

Take Peter Boyce's influential book *Human Factors in Lighting* (2014) for instance: summarizing more than a century of knowledge on light, the book explores the relationship between light and humans. Substantial knowledge about light is presented, and yet there is an omission of any mention of social or cultural aspects of lighting – as if cultural traditions and social life are unknowable or secondary to the individual bodily response to light, rather than embedded in the very way people use, sense and make sense of light. What has emerged within lighting design and engineering is, as Boyce also laments (2017), a gap between 'the art of light' – dealing with the aesthetics and the end user, albeit at a more intuitive level – and 'the science of light' that deals with developing new technologies and understandings based on quantitative measures. The latter seems to reflect a scientism where the main universalistic assumption is that 'good lighting' resembles the sun, and consequently 'daylight' has been the outset for optimum colour rendering in light bulbs; lighting in this sense is understood in terms of its impact on a human body, with technical efficiency and quantitative measurements as an end-goal (Boyce 2017). This establishes a dehumanizing separation of the human *body* from the human *person* in what may be termed a 'lumination gaze' (cf. Foucault 1973). I argue that this version of 'human-centred lighting' is at risk of being an *asocial perspective* on both light and what it means to be human – even if this is not the intention.

In part, this omission is a result of the still relatively scarce literature on the social and cultural aspects as I highlight in Chapter 2. Broadly speaking, most research on the relationship between lighting and humans has been aimed at cause–effect research: if you lower light level to X, then Y will be the effect. Yet there are other kinds of research and knowledge, for instance one that aims at conceptual development for understanding what is going on and why. This book is meant as a contribution to a qualitative understanding of how humans adapt to, adopt, and live with light. It offers an exploration of the meanings, premises and uses of light in domestic lives. This is not intended to be an experiment with a 'before and after' to see the effect, but towards understanding what people do with lumination, and why. The premise of such a 'social' approach is that research on lighting practices and atmospheres needs to be situated, and understood, in the particular context, where concepts and insights can guide our understandings of other contexts.

An awareness of the sociality of light is particularly important when considering climate change. New technologies may be less polluting or afford a better quality of light, but if they do not fit into people's cultural ideas and norms of how a place should feel, or even the relation between light and the objects it illuminates, their success may be limited. Beyond their technological efficiency, their capacity to offer change rests on their ability to make sense, and this sense-making, I argue, is also about shaping atmospheric spaces. Cecilia, a visually impaired person with strong sentiments about energy-saving light bulbs, illustrates this point, even if she did end up using them: the energy-saving light bulbs stood out as a separate entity of the home, rather than being integral to the domestic *lumitopia.* Part of the reluctance to accept the energy-saving light bulbs concerns, of course, their current quality, price and lack of standardization. But there is also a narrative that energy-saving light bulbs are of a lesser quality than incandescent ones, regardless of the validity of such claim.

The transition to energy-saving light bulbs has made the quality and quantity of light and lighting a hot topic in Denmark and elsewhere. Rather than a technical exploration, this book elucidates how quantity and quality of light unfolds through practices and atmospheres in domestic spaces in Denmark; that is, the cultural fit of a technological fix to energy consumption. The aim is to achieve a broader understanding of the role of light in social life through the power of example of a case study of lighting in Denmark. It aims to illustrate how ideas about the good life, home, morality, and identity are not only subjective mental capacities, but are achieved through technologies. Previous books have shown similar issues in the historical transition to electrical light (Barnaby 2016; Garnert 1993; Nye 1990; Schivelbusch 1988), yet still little is known about light as it is embedded in social life in contemporary society (but see Edensor 2017) and the cultural horizons it is part of (cf. Bille & Sørensen 2007; Daniels 2015; Kumar 2015; Wilhite et al. 1996; Winther 2008).

The central argument in this book is that, beyond offering visibility, lighting *is a practice of attuning atmospheres* for a variety of activities to take place in the home, such as social gatherings, cooking, relaxing, entertaining, cleaning, and feeling secure. While much research on light has focused on how much and what kind of light is appropriate for certain practices – such as reading or working more efficiently – I wish to reframe the question into an atmospheric perspective, and instead explore the atmospheres people focus on, in which such practices are embedded, or help to shape. For instance, the best 'reading light' has been investigated by studying how well people of different ages are able to read a text under a certain light in terms of speed, quality or fatigue. An atmospheric approach will emphasize how 'reading' may be embedded in different atmospheres depending on the type of text and situation, where candlelight, or lighting a distant corner or adjacent room, are more relevant luminous landscapes to understand than cognitive reading speed, quality or fatigue. This shifts the focus from the *body* to a *spatially felt* object of study. It is not a shift from body to space, but rather to the affective and emotional embedded body *within* space. And in this case that space is the home. From the perspective of this book, dwelling and home-making are, in essence, a search for – but not necessarily succeeding in achieving – the right atmosphere to express and shape the embodied practices and meaning of its users. Although many nuances naturally apply, to reduce the point of the book to a single phrase, try to think about lighting in these terms: people do not only buy a lamp or light bulb to offer visibility; they buy the technical means for (hoped for) atmospheres to unfold. Such atmospheres may end up homey, scary, relaxing, intense, intimate, or alienating. What they are, is an empirical question; the difficult part is to frame questions about lighting in atmospheric ways.

The example of Cecilia – a visually impaired person orchestrating light and atmosphere to meet social ends while scorning energy-saving light bulbs – raises the central questions addressed in this book: What is good lighting (not), for whom and why? What are the social consequences and cultural premises of lighting practices? How does lighting take part in shaping the atmosphere of a place? And what is an atmosphere in the first place – a cosy one in particular? In essence, this book is about the cultural logics and practices of light that people make use of, or are confronted with, when affectively attuning spaces, themselves and others through the atmosphere they are part of. The question of how to light 'the right way' and make the atmosphere 'feel right' leads to the central analytical point in this book, namely that light is not shed *on* the home but *for* the home (see also Bille & Sørensen 2007; following Bolt 2000). The distinction marks both a theoretical and a methodological difference: 'light shed *on*' separates technology from human life, objectifying both on its way, which leads to questions of causal effect. 'Light *for*', on the other hand, highlights how technology is intrinsically embedded in human perceptions, practices and meaning-making, which I propose essentially leads to a more human-relevant approach in which the relationship between the sentient embodied subject and the world emerges. People thus live *with* light and not just *in* it. Importantly, this is not about presenting people as living in harmony and balance with light, but about attributing light an active role in people's lives, for better or worse. It is about understanding the different relationship emerging between humans, space, and light sources that at times may stand out as separate entities, and at other times as indistinguishable from oneself and one's surroundings. Light *on* promotes a visual dominance that can be represented in a picture: something people look or gaze *at*. For instance, studies have placed participants in a magnetic resonance imaging scanner showing a picture of a room, and then scanning their brain to see what areas are activated by this passive *looking at* (e.g. Vartanian et al. 2013). A photographic representation of spaces, however, runs the risk of stripping a space of its embracing, sensuous and corporeal essence, making it unreal and untouchable (Pallasmaa 2005: 32). Through vision people may achieve a closeness – being *at* a place – but that is not similar to being *in* a place.

Light and lighting practices offer a (potentially) transformative attunement of people: making people feel a certain way, letting them influence or be captured by an atmosphere. Thereby atmospheres and lighting are also ways of establishing and confirming cultural norms of what to feel and what to do. People learn, with all their senses from early on how a home smells, looks, and feels. They may not like it, or disassociate themselves from it later on, but the atmospheric imprint is there nonetheless. There may naturally be an imbalance between ideal and achieved outcomes, or contesting ideas and practices, as well as issues of who the agent is in orchestrating the light and the counter-measures taken. Furthermore, light is not the only technology used to tinge atmospheres, as the central premise of atmospheres, as explained later, is a multi-sensuous co-presence of subject(s) and objects. But light is an important aspect, as will be illustrated. Nonetheless, the embodied effort to attune oneself and one's surroundings through material objects, even intangible ones like light and sound, is fundamental to understanding social life, particularly as it unfolds at home, which connects light to cultural registers such as power, gender, identity, morality, and comfort.

The material presented in this book originates from fieldwork in Copenhagen, but the point is not to explore a particular 'Copenhagen light', or

FIGURE 1.2 *Atmospheric 'living light'. Photo by the author.*

wider issues about Danish or Nordic culture as such (see Gullestad 1984; Jenkins 2012). There may be different social effects and practices across Denmark, for instance between living in an apartment, a low-energy house and a rural farm area, during events in the public sphere, or between social segments or generations. However, judging from the dominant place of the term *hygge* and *hyggelys* (cosy light) in both private and public spaces in explanations of Danish culture, the emphasis on using light to shape such atmosphere is general in Denmark.

A common critique of case studies is that they do not allow for generalizations (discussed in Flyvbjerg 2006). Yet by keeping a focus on lighting practices and asking 'what good light is' – and going into depth in that somewhat normative question – this book offers qualitative insights into how light and lighting is used to meet particular social, emotional, at times even political, ends, through people's everyday practices. Such use, effect and meanings of light may take different shape elsewhere, but the main point is to highlight how a focus on atmosphere and light is at the heart of people's everyday lives, regardless of whether the point is informality and cosiness, as in Denmark, or formality and protection, as in a case study I did in Jordan (Bille 2017), discussed briefly in Chapter 6. Approaching social worlds by focusing on the atmospheres such worlds unfold through, may help us better understand not just *what* is happening in people's homes and *why*, but also *how* homes are shaped through emotional and sensuous qualities of lighting technologies. This leads us to the next central question: what *is* an atmosphere?

Atmosphere

The term 'atmosphere' is used in at least two very different ways in academic literature (Gandy 2017a; Hasse 2014: 215; Henckmann 2007: 48; McCormack 2008: 413; cf. Olwig 2011). First, it denotes a physical phenomenon surrounding a planet or a star as a layer of gases, in what may be called a *climatic understanding*, including weather and environment. Second, there has been a metaphorical adoption of the term, where atmosphere is understood as the affective, sensuous experience of a place in what may be called an *existential understanding*. This latter version addresses how a question about what it means to be human is also fundamentally concerned with being sensuously affected *by* and *in* places, and the process of affecting such places. This existential version has been discussed particularly in philosophy (Böhme 1995, 1998a; Griffero 2014; Tellenbach 1968), but also more recently in both social science (Albertsen 1999, 2012; Anderson 2009; Anderson 2014; Bille et al. 2015; Dupont & Liberg 2008; Edensor 2012; Hasse 2014; McCormack 2008;

Philippopoulos-Mihalopoulos 2013, 2016), and in research on tools for promoting consumption (Custers et al. 2010; Healy 2012; Kotler 1973; Schüll 2012; Sharma & Stafford 2000; Turley & Milliman 2000). However, these two kinds of atmosphere should not *a priori* be separated. For instance, changes in the climatic atmosphere affect the existential atmosphere by calling for more energy-efficient technologies, such as new light bulbs, thus also changing the everyday existential atmosphere by offering more 'cold' light than the incandescent bulb. Furthermore, while the weather may be seen as part of a climatic atmosphere, it is also closely entangled with human perceptions of places, such as when a rain shower suddenly transforms a trip to a sunny beach (Bille & Sørensen 2016a; Gandy 2017a; Ingold 2007a, 2012). Also, the very distinction between a climatic and an existential version is a highly cultural distinction to make in the first place since, for instance, the Arabic word for mood, atmosphere and weather is the same – *jow* – as illustrated in the common greeting '*Kiif al-jow?*' ('How is the mood/weather?') (see also Daniels 2015 on Japanese terms).

One of the most prominent scholars highlighting atmosphere as a fundamental aspect of being human is German phenomenological philosopher, Gernot Böhme (cf. 1993, 1998b, 2003, 2005, 2006a, 2013a, 2017). As discussed in Chapter 4, Böhme explores the nature of atmospheres as they unfold in terms of both the undoubted familiarity people may have with a place, as well as how a place may overwhelm a person and seize them emotionally or vice versa. In its most essential form, Böhme conceptualizes atmospheres as the co-presence of subject and object (1998b: 114). This means that one will look in vain if one tries to point out its location or material form – its nature of being (its ontology) – as somehow reduced to the material surroundings, or solely in the mind of a person. This would be to ask the wrong question. Instead, they have to be felt by a sentient being as a 'subjective fact'. They *exist*, although in a different way from, say, a chair. But that does not make them any less real: they are 'half-things' (1998b: 114, 2001a: 59; Schmitz 2014: 39). In an oft-cited paragraph, Böhme exemplifies how:

> atmospheres are indeterminate above all as regards their ontological status. We are not sure whether we should attribute them to the objects or environments from which they proceed or to the subjects who experience them. We are also unsure where they are. They seem to fill the space with a certain tone or feeling like a haze.
>
> 1993: 114

Atmospheres can thus be conceived of as 'tempered spaces' (Bollnow 1963: 230), where subject, object and space are inseparable from each other, aiming to completely disregard the separation to begin with. In that sense, I have argued (2015a) that despite the possibility of staging atmospheres through material strategies, they should not be seen to emerge *in* the relation between subject and object but rather unfold *as* the relation; they are always there, it is their character that changes. This challenges traditional views of feelings, for example, by highlighting that they are not necessarily 'located' in the mind but are 'out there' as the relationship shaped among people and materialities. Atmospheres in this sense describe 'a feeling whose peculiarity is that of being spatialized' (Griffero 2014: 101). While it may work in broad terms to conceive of an atmosphere as the 'co-presence of subject and object', the symmetry between the subject and object, with equal proportions of each, is something that should be empirically grounded, so should its temporality. People anticipate what a space *should* feel like, and the mood of a person also affects the way s/he perceives the world. But people are also affected by other people's moods (e.g. Brennan 2004), even if no wholesale transmission of affect occurs, along with the affective presence of material objects that may potentially seize us (Armstrong 1971). Jürgen Hasse, who has been seminal in locating the philosophical discussions in nuanced empirical case studies, calls this potentiality the 'affective overprinting by an atmospheric impression' (2014: 220). 'Potentially' is an important nuance here because it is not certain that the atmosphere will capture the people it embraces or permeates. A dinner party may be staged to be romantic or cosy, with music, wine and lights, but this intended atmosphere may not capture everyone in the room. The host may even notice this and attempt to adjust the atmosphere by changing the material setup, the conversation, light, music or other elements that make up the whole.

The experience of atmospheres is not about exploring causality but about emergence. As an atmosphere comprises different affective elements of people and material phenomena – such as light, sound, dresses, walls, chairs, etc. – they are also unfolding on different scales of temporality. As a consequence, perception is not just about what *I* feel or 'see', but how affective presence is momentarily dispersed in spaces as a 'tone' (e.g. Böhme 2017: 74). This 'tone' may then be qualified in different ways, such as 'festive', 'serene', 'uplifting', or, as Cecilia puts it, '*hygge*'. Such cultural terminology reflects both what is being sensed, and what should be sensed: the atmosphere *should* be cosy in Cecilia's home. Yet the qualification does not imply that atmospheres are always positive sensations, as post-terror attack sites, hospitals and graveyards may equally produce eerie atmospheres (Adey 2014; Adey & Anderson 2012; Anderson 2010; Anderson & Adey 2011). '*What* they are' and '*that* they are' are thus two different questions. Although recognizing such cultural understanding, Böhme rarely delves into the nature of what atmospheres *should* be, and this poses a problem, because it highlights that a phenomenological account often in the end comes out ahistorical or without a context (see Bille 2015b; Daniels 2015; Olesen 2010;

Pennartz 1986). Atmosphere, in a more anthropological sense, is about how concepts and the human sensorium are validated and shaped through socio-material processes. The objective of this book is thus to offer an understanding of such atmospheric premises and processes when exploring light and lighting both broadly speaking, and in the particular material organization of homes and homeliness in Denmark.

Embraced with light

Tim Flohr Sørensen and I (2007) have argued for increased attention to the question of what good lighting is by exploring how light, shadow, colour and darkness embrace human lives. While the article was by no means the first one to show that light had mattered, historically, in social lives (Baxendall 1995; Blumenberg 1993; Jay 1993; Kapstein 2004a; Zajonc 1993, and many others), there had been very few systematic and dedicated studies (such as Garnert 1993, 1997; Wilhite et al. 1996) in the social sciences and humanities to how contemporary society is shaped by landscapes of light, darkness, shadows, glare, luminance and illumination (recent studies include Cross 2013; Ebbensgaard 2015; Edensor 2012, 2017; Edensor & Millington 2009; Hauge 2015; Isenstadt et al. 2015; Jensen 2014; Kumar 2015; Meier et al. 2014; Morris 2011; Shaw 2015; Stidsen et al. 2014; Thornton 2015). Understanding lighting in this perspective is about the way both natural and artificial light, most often in combination, is set up to promote social and affective occurrences: the meanings and presences light and darkness give to material surroundings. Natural and artificial light play a central role in everything from movement, consumption and work in public spaces to intimacy and relaxation in the home; from the meanings and symbolism light carries as metaphors for truth and spirituality, to the changes it makes to the presence of other material objects. Light is part of the way people sense and make sense of the world, and it is designed and placed to achieve socio-affective, political and economic ends.

While the impact of technologies in human lives has not always been at the centre of attention in social science, there was a rapid spread of Material Culture Studies and Science and Technologies Studies at the end of the 1990s that explored the social life of things, the meanings and practices they take part in, and how knowledge and technologies are shaped by social worlds (e.g. Appadurai 1986; Bijker 1995; Latour 2005; Law 2002; Miller 2005; Mol 2002; Shove 2003; Shove & Pantzar 2005). As illustrated above with Cecilia there is obviously a materiality to light and lighting in terms of the infrastructure it takes to shed light through electrical cables, contacts, lamps, dimmers and bulbs. Yet, light is intangible, and material culture studies have predominantly dealt with the present and tangible objects. This, however, raises a concern that omitting the intangible or absent in studies of materiality is to see only half the picture of the role of materiality in human life. It is, I argue, a mistake to equate materiality with tangibility. In the same way as there may be a materiality *in* and *of* absence (Bille et al. 2010), there is a materiality to light in the qualities of measurable matter, but more importantly here in that it exposes textures and surfaces to human perception, and offers sensations of brightness, luminance, colour reproduction and temperature (both in terms of colour and thermal). Addressing the materiality of light is to address both the object of light itself as electromagnetic radiation (of which only a portion offers visibility to humans), the sensation of it, the meanings it acquires, and the social life it helps shape (for a discussion on materiality see Ingold 2007b; Miller 2007; Tilley 2007).

Particularly relevant to this book is thus also the role of material culture in understanding people's sense of home. In such a perspective, an important focus is on the sensuous and emotional aspects highlighting how a house may be seen as an object, while a home is a process of appropriation imbued in sensory and emotional histories and habits (cf. the journal *Home Cultures*, and Blunt & Dowling 2006; Buchli 1999, 2013; Cieraad 1999; Gullestad 2001; McCracken 2005; Miller 2001; Pink 2004). 'Home is not this house, but this atmosphere', as the Danish band Bellhound Choir sings on their song *Home*. The intensity of such atmosphere may by noticeable, overwhelming, or indiscernible, yet in each case it is there. This does not necessarily entail harmony and joy, as the home may also be a place of violence, the everyday life of broken dreams, or encompass the lack of skill, will or money to shape a balance between aesthetic ideals and reality. As one of my informants, Henrik, notes, 'a home is never finished'. There may always be things that the inhabitant is not quite satisfied with, and often that is part of the point of making a home: that there are yet things to be done. Studies of home-making also show how small objects like food or memorabilia may create a sense of home in places of temporary residence, or how the home comes into being, is engendered, moralized and emotionally saturated through shopping, practices and sensuous experiences. In essence, material culture holds a central place in shaping homes, sustaining moral regimes, and forging identities, but what is needed is to expand such focus on materiality to put more emphasis on the intangible and ephemeral: what is a home and the sense of home if not a feeling and continuous process of relating through materiality? The present study thus adds to this growing literature on the materiality of the home by focusing on the practices of shaping homely atmospheres through material culture – lighting in particular.

Lighting practices

Beyond being a common atmospheric element in the meanings and presences of the world, light is also entangled in domestic practices. Turning on the light when going to the toilet or into the kitchen often happens as the most taken-for-granted activity today. Many of the informants[1] presented in this book had rarely, until asked, thought about what they do when they come home and why, habitually turning on lights, and forgetting whether they turned them off again, only sensing it at the margins of attention. Light is mostly taken for granted as a part of ways of seeing; unless there is a power cut (Nye 2010), one may forget that no more than a century ago most of the world was not lit by the flick of a switch, but by firing up matches, ignition devices, a gas or oil lamp, or lighting a fire. Back then, it involved some sort of materials, skills, preparation and attention to turn on the light; if not to avoid fire hazard then to avoid consuming expensive resources. In that sense, with those technologies the materiality of light and the perception of danger may not be seen as separate characteristics, but rather as being *embedded in* the technology (Nye 1990; Schivelbusch 1988).

Lighting is as shown above central to the visual shaping of the home. But it is not only about vision or light. It seems as if a defining characteristic of genuinely embodying a place is whether one knows 'by heart' where the light switches are. Switches are mostly located by doors, thus equating light with physical presence. Yet in Denmark one often finds lights in places with no people, or, conversely, very little light if people are present.

Today, however, artificial light is most often shed *on* us without our influence, rather than *by* us. As a consequence, we may find buildings with large glass façades allowing for natural light to spread – although slightly tinted by the glass – into a space that is simultaneously lit by artificial light. It may be fully acceptable to sharply distinguish between natural and artificial light from a scientific or architectural point of view. Obviously, there is the distinctive natural light from the sun and the moon, which architects mould with walls and fenestrations. A toilet or cellar may have no natural light entering, and artificial lighting may have no role in a garden in the daytime. Yet, from an anthropological and phenomenological point of view of understanding lived experience in places, being in a place will often entail both – even if just a traffic light or shop sign in neon – making it more relevant to ask when a separation of natural and artificial lighting makes sense, rather than have such separation as starting point.

Built spaces are standardized as to amounts of luminosity required to secure productivity, energy efficiency and comfort, recently by automatic dimming or brightening artificial light as the clouds block the sun to sustain a certain amount of *lux* (a term used about the quantity of light – *lumen* – in a square metre). Even more recently the move towards dynamic or circadian lighting – where lighting follows the daily rhythm with more bluish light in the morning and more reddish light in the evening – is helping people to cope with the lack of sunlight they are exposed to by increasingly spending more time indoors. Furthermore, public indoor spaces have increasingly implemented automatic lighting as people enter or leave a building (or simply deactivated it if a person stands or sits still for too long), regardless of whether it is possible to see without the light on or not. There is in this sense a 'scientification' of lighting taking place, in which the user is a physically responding *body* more than a social *person*. According to Claudia Dutson, control is thereby 'taken out of the hands of the users and given to the services of the engineer, who it is assumed knows (objectively) best and can overrule dangerous things such as personal (subjective) opinion and preference' (Dutson 2010: 42).

As much social science has shown, it may however not always be fruitful to make the individual body or person the starting point for analysis. Lighting norms and visual materializations of ideas, experience and knowledge, for instance, must also be understood beyond the attitudes, behaviours and choices of such individuals (Shove 2003, 2010; Shove & Walker 2014). These individual choices are entangled in technological systems of electrical wiring and available resources – of bulbs and electricians; of climate politics and standardization and the sciences informing them; of architecture and designed surfaces; and, most importantly, of the meanings and social distinctions achieved through lighting – all of which are grounded in a particular cultural context. We may speak of 'light cultures' as a mixture of individual sensuous responses, preferences and actions that are somehow shared in *kind* among a group of people despite diversity in *degree* (or at least shaped in reference to such groups). It is the outcome of socio-technical systems of standards of electric infrastructure and wires. It is embedded in the traditions of architects and designers of moulding light and lighting. It is in the politics and mundane ways of talking and thinking about light. It is an anthropological notion of culture that does not entail fixed boundaries of 'this is what the Danes are like' (although with *hygge* there clearly is an element of national identity involved, as discussed in Chapter 5) but centres more around the emergent and relational practices, perceptions and meanings evolving through socio-material interactions. Seen from such a social perspective of what people do, there is then a 'contemporaneity' where light all at once is a marker, an action instigator, a sign and a stage-setter, which turns a *physical space* into a *social space* (Ardener 1992: 27). Materiality, which in the case of light is intangible, becomes the means of making human spaces. The key point is that people may be informed by the newest knowledge about the physiology of the human body or light – making sure to get sunlight or avoid the particle pollution from candlelight – but they are also informed by social and cultural norms about what the world *should* look like, ideas about the good life, class, morality, gender and power, irrespective of the physiological soundness or possibilities of that choice. This goes to show that people use light to do many other things than simply to see, and the introduction of new technologies, or persistently holding on to old ones, helps to show how social and cultural ideas and ideals are integral to understanding how people use and understand light as part of informing social life.

Light and lighting

One way of approaching light is to differentiate between 'light' (a material form emanating from a natural or artificial source reflecting, changing or absorbing on material surfaces and an object of scientific scrutiny), and 'lighting' (as a practice of manipulating, designing and appreciating luminosity in itself or to inform other practices). To focus on lighting is to focus on the relational occurrences where light, as a material phenomenon, comes to matter to people. Asking 'what good lighting is' as seen from an anthropological perspective rather than from a technological, biological or architectural one, is also a very normative question, although that may not entail that the answer is normative. Engineers, psychologists, architects and medical science have shown how much light is optimal for a task, how to construct new technologies and aesthetic forms, and how lighting affects cognitive skills and physiological needs. The influence and importance of such studies should not be disregarded, even if it must also be recognized that they most often rest on assumptions that are culturally, or historically, contingent.

Yet the point of social studies of light and lighting is to look at what people actually *do*, what they say they do or know, and analysing and extracting the premises of such actions, knowledge and values by tapping into the lived experiences. Just because the optimal light for reading may potentially be accurately measured according to age and sight of a human body, there is nothing to suggest that the 'practice of reading' is not bundled up with other practices, such as attuning a cosy atmosphere, which may encompass other types of lighting preference. As will be shown throughout the book, the results of such an approach offer insights into the social and cultural aspects of lighting that have been largely neglected, and highlight more theoretical and analytical concerns about human interaction with things through the concept of atmospheres. This approach then offers the potential for a better understanding of what kind of light and lighting is used in various contexts, a more flexible and interactional vision for future lighting technologies, and understanding that technologies originate from, and are embedded in, social worlds, rather than somehow external to them.

In essence then, three central themes emerge from the book: 1) to approach homes, and spaces more generally, as a practice of attuning atmospheres, investigating the agents, the conflicts and the premises of such atmospheres; 2) to offer a cultural and social grounding of the phenomenology of atmospheres, exploring the skilled visions and values that are embedded in lighting as atmospheric element, and 3) to reframe questions about lighting towards social interactions beyond the individual body. In essence, seeing light as something people live *with*.

Many ways to study light cultures

Together with assistants, most notably anthropologist Matilde Lykke, I conducted around 70 formal semi-structured interviews lasting from 20 to 120 minutes, following an adapted interview guide. Unless otherwise stated, all quotes in this book are transcribed from audio recordings. Anthropological data are, however, not only generated when the recorder is on; they also in large part emerge when you read news and reports, attend small talk at lighting conferences, or are simply invited to dinner and listen in on people's talk about lighting and atmosphere. As stated earlier, most social situations where I talk about my research topic end up with people telling me about their practices and experiences with light. While it is tempting to suggest that the formally recorded interviews are more valid than the informal interviews and conversations, I consider the insights they offer to be of equal importance, although I only quote recorded interviews.

Furthermore, there is more to qualitative methods than formal and informal interviews and conversations, illustrated, in particular, by the esteemed position of participant observation in anthropology. The research presented in this book is based on visiting the informants, rather than living with, or following them for a whole day, month, or year. Yet, I did 'settle in' on Islands Brygge, Copenhagen, living in the same type of apartment as some of the informants for two years, thus being able to watch what people did, follow their routines as they got up in the morning or came home during weekdays and weekends. I also participated in setting up the extensive Christmas lights explored in Chapter 7, spending a day in trees, on roofs and in the garden while talking to the informants about their practices and the objects, and eventually returning for the 'turning on the lights' event for several years in a row. Informants were found, with a few exceptions, by knocking on doors and 'snowballing' from there (asking if they knew someone who would be interested in participating). In essence, the methods are more or less standard qualitative methods with the addition of cultural probes to initiate discussions with informants (Brinkman & Kvale 2014; DeWalt & DeWalt 2011; Pink 2009; Spradley 1980).

Very different types of living conditions are present in the study, also in terms of owner conditions: one-family houses, refurbished old military barracks, early twentieth-century style apartments, early twenty-first-century apartments, and mid-twentieth-century social welfare housing. Most interviews were conducted on Islands Brygge, explored in Chapters 3, 4 and 5, while interviews from Urban Planen are the focus of Chapter 6. The case study of the Christmas lights in Chapter 7 is from a residential area in Tårnby, a Copenhagen suburb. Interviews were also conducted in houses in another Copenhagen suburb, Hvidovre, and two visually impaired informants were interviewed in the greater Copenhagen area. The informants were all over 18, with most aged between 25 and 45. There are slightly more women among the informants, and very much so in terms of whom I have ended up actually quoting. Women were more often the everyday caretakers of the home, and thus also largely responsible for interior design, but it must also be noted that male informants expressed same general sentiments about lighting practices and qualities, and I have not been able to systematically find areas where men and women have markedly different practices. The informants range from students, public clerks, CEOs and private entrepreneurs, to people on social welfare. Almost all of the homes were photographed, mostly for documentary reasons, but are presented in this book for illustration purposes also.

What to expect?

The reader must be warned at this stage that this is not intended to be an anthropological monograph of lighting in Denmark, where we get to know people like Cecilia in depth throughout the book. Instead, the book intentionally moves from broad introductory chapters on social perspectives on lighting and conceptual discussions of atmospheres in Chapters 2, 3 and 4, to more ethnographically driven chapters from 5 to 7 and finally end with broader reflections on politics of lighting in Chapter 8. At times it offers broad reviews, at other times it is ethnographic, anecdotal or even polemic. Despite this diversity in scope, the chapters in different ways share the focus on the importance of everyday homely lighting practices, and the cultural appreciation and social life that light is embedded in.

The first step in advancing a more social understanding of light, in line with how material culture studies have highlighted the role of things in human life, as discussed above, is to explore selected studies on the social effects, practices, and meanings of light. Chapter 2 – The sociality of lighting – outlines the expanding literature on social aspects of lighting, particularly in themes that have dominated lighting research, such as health, energy consumption and security. While some themes, such as 'security', have defined public lighting since early times, 'health' increasingly became an issue during the nineteenth century, and 'energy consumption' during the latter half of the twentieth century. The chapter outlines insights from different social and cultural approaches to lighting in order to complement the extensive technical and medical knowledge of light and show how the way people shed light *for* the world informs ideas about morality, gender, security, well-being, etc.

From *outlining* selected social perspectives on light and lighting, the analysis turns to the atmospheric aspects of light in Denmark. Chapter 3 – In the vagueness of Hygge – is a more *conceptual* exploration of one of the central atmospheres underlying social life in Denmark, namely cosiness (*hygge*). The chapter discusses how there is a profound lack of addressing vagueness and messiness as a premise in an academic understanding of light in everyday human life. Through ethnographic examples, and methodological and theoretical discussions, it seeks to clarify the notion of *hygge* as an example of this vagueness. By doing so the chapter highlights the importance of scrutinizing atmospheric descriptions in detailed ways.

In line with this conceptual clarification, Chapter 4 – Atmospheric realities – discusses the recent literature on atmospheres. While Böhme mentions the importance of culture in understandings of atmospheres, his project is philosophical. Expanding upon the previous chapter's outline of a particular *kind* of atmosphere, this chapter discusses atmospheres as not merely about a momentary perception of the world, but embedded in practices that allow us to shift focus from the ontology of the individual object to the practices and ontology of elements.

From these broad conceptual introductions, the book now turns towards the *practices* of lighting and how light is part of structuring social lives. Chapter 5 – Atmospheric communities – explores the mundane daily life of lighting in detail by highlighting how light plays a central part in how people understand the neighbourhood they are part of, and the feelings and boundaries of home. That is, light helps shape people's sense of community by articulating the boundary between spaces, thereby highlighting the shared meanings and atmospheric competences through which light unfolds. It thereby emphasizes how belonging and sharing identity may be both a sensuous and atmospheric occurrence. From people cuddling up in a 'reading cave', to staged social gatherings, to notions of Nordic lighting, the chapter illustrates how ideas about boundaries shape communities of things, people and places, through light.

While the dimmed lighting of a *hygge* atmosphere is widespread in Denmark, it is of course not the only way of illuminating spaces. Chapter 6 – Qualifying bright lighting – homes in on diversity, and explores the many *meanings* light has. The chapter highlights bright light as a contrasting phenomenon to the dimmed *hyggelys*, and emphasizes how different cultural meanings and appreciations of light are embedded in such lighting. The starting point is a social housing with a high percentage of immigrants, and through examples it shows how lighting comes to shape ideas about prestige, social gatherings and notions of privacy. Here, light becomes a signifier of diversity, object for imagining future lives, and, at times, even state of mind. As the ethnographic descriptions from the households reveal, some Danes with immigrant background also appear to adjust their lighting culture somewhat, leading to an investigation of continuity and change in cultural lighting practices. The importance of understanding the premises of lighting norms is further illustrated by an example from Jordan on the role of bright light and new lighting technologies, which highlights the social meanings embedded in (changing) lightscapes.

Chapter 7 – Ignorance and illumination – takes a closer look at how practices of attuning atmospheres through lighting are shaped by people's knowledge; that is, it moves from the sensuous and affective aspects dealt with in the earlier chapters, and turns to the *knowledge* it takes to operate and interact with lighting technologies to shape domestic atmospheres. The new lighting technologies have forced new ways of thinking about light. Yet these technologies have also been met by resistance in some cases, and have led to many wrong purchases and consumer frustration, which we explore through a case study of selecting the right bulb. But in other cases, the new technologies have allowed for skilful technical knowledge to couple with atmospheric desires. This chapter also takes a closer look at a case of extensive lighting to create a Christmas atmosphere, often scorned as a waste of energy and competitive materialism, but which is also closely tied into Christmas messages of joy, altruism and emotion. Emotions, however, may not always be productive. By following instances of how *lack of knowledge* leads to frustration and ridicule, the chapter explores the notion of the technological ignorant, the *technorant*, to understand how knowledge and social positioning are part of atmospheric attunement. Light, and knowledge about light, become more than simply a cognitive and sensuous matter, but also an emotional and atmospheric one, spanning from the frustration elicited when one buys the wrong bulb for *hygge* to the joy of Christmas atmospheres.

The last step in the analysis of the social life of lighting and atmospheres is exploring the *politics* of lighting, materializing moral issues in the age of climate technologies. Chapter 8 – Reflections of sensory politics – looks at ideas of pollution and how technological fixes to curb demand are materialized in energy-saving light bulbs. The chapter argues that technology is neither good nor bad; nor is it neutral, since it mediates moral claims most evidently seen in the energy-saving light bulb and the embodied practices of turning lights on and off. In this way, the chapter concludes by looking at how lighting practices and atmospheres are entangled in broader political issues of materializing morality, by tying the need to care for the climatic atmosphere together with the light used to orchestrate existential atmospheres. The chapter completes the book by reflecting upon how people's domestic lighting centres on attuning people and spaces to the task at hand. It is about shaping and being shaped by the atmospheres that light, among many other material forms and practices, helps constitute. People produce atmospheres, sense of self and social positions through ideas about 'good lighting', which, in turn, require energy. Through such a lens we may understand how human perceptions and practices are central to the efforts to cope with climate change, and thus also centrally place the human *person*, and not just *body*, in the understanding and orchestration of light and lighting.

2

The sociality of lighting

Introduction

The year 2012 was important in the history of lighting in Denmark. This was the year where the last of the incandescent light bulbs were banned from production and import, as part of a European Commission eco-design directive to lower energy consumption. It started with 100W and frosted glass bulbs in 2009, followed by 75W in 2010, 60W in 2011, and 40/25W in September 2012. The remaining stock can still be found online, or in the homes of those who hoarded them in defiance of the quality of the energy-saving light bulbs. 2012 effectively marked the end of large-scale lighting by incandescent bulbs. Few other occasions in world history have resulted in the same kind of politically sanctioned transition from one type of technology to the next, where the replacement technology – the compact fluorescent light bulb (CFL) – is likely to be very short-lived. This is not only because the CFL bulb's lighting quality is technically poorer than the one it replaced on several parameters, but also because an even better technology is already being rapidly developed: the light-emitting diode (LED). Even while publishing this book, one is bypassed by new lighting features and technologies, such as Wi-Fi and Bluetooth-controlled LED light in millions of colours. The ban was enforced while a mitigating hope of acceptable visual comfort was promoted in commercials and information campaigns by claiming that the light is not as bad as rumours have it, or by pointing nebulously towards the next generation of LED bulbs.

The phase-out of incandescent light bulbs has been enforced in Asia, North America and Europe, and with good reason. In Denmark, it is estimated that around 13 per cent of household energy is currently used on lighting,[1] although one could argue that in terms of climate debates around CO_2 emission, the problem is not energy consumption, as such, but how that energy is produced.

To solve the problem of increasing energy consumption of electrical light and the use of fossil fuel to produce the energy, at least two solutions are ready at hand if approached from a technical perspective: developing more effective light bulbs and deriving more energy from renewable resources. The incandescent light bulb ban was the result of the former, and with the rapid development of new lighting technologies we are now seeing changes in the visual orchestration around the world, from cheap but efficient solar lamps in third-world countries, to innovative commercial and aesthetic lighting in urban centres. Beyond the light bulb, light sensors installed in offices during the last decade automatically switch on and off as they detect movement, and public street lighting is refitted with energy-saving bulbs at times staged like theatre scenes. Like in the past, architecture also plays a crucial role, as with the oil crisis in the 1970s when only small window sections were allowed to save heat emission (but which also led to more electrical light being used), or, as window quality is now improving, by having larger window areas, allowing for more light and natural heating to occur (and extensive energy consumption on cooling in the summer).

Consumer adoption of energy-saving light bulbs was, however, not as rapid as politicians may have envisioned in Denmark. The ban was met with criticism by the general public and parts of the scientific community due to the poor quality of energy-saving bulbs on sale in supermarkets, where 51 per cent of the population in 2016 bought their bulbs.[2] A large part of the cause was lack of standards as well as 'Geopolitical dimensions to trade negotiations, corporate power dynamics, strategic state interventions in the fields of research and development, and local sites of policy experimentation' (Gandy 2017b: 1099). In such a void the supermarkets have thus also adjusted their assortment, often focusing on the less energy-efficient halogen bulbs than the LED or Compact Fluorescent bulbs. Furthermore, several other issues came up, such as less heat emission from energy-saving light bulbs, leading people to compensate by heating more; there is a quality reduction in the spectrum of light from the energy-saving bulb; the compact fluorescent light bulbs have mercury in them, making them more polluting and requiring people to adopt proper disposal practices; they are heavier and bigger and so cost more to transport, not to mention the high price of purchase may put people off, even if they last longer. Finally there is also the so-called Jevon's paradox or 'rebound effect' whereby the efficiency increase in technology is followed by an increase in demand. In other words, as the light bulb becomes more efficient it is used more often than its less efficient predecessor, therefore lowering the energy gain (cf. Frondel & Lohmann 2011; Gram-Hanssen 2013; Horta et al. 2014; Howarth & Rosenow 2014; see also the review in Jensen 2014).

I will call the above a 'light as energy' approach, where engineers have a central role in the solution. However, this approach is also increasingly nuanced by social scientists who show how a social approach to light as energy is central in understanding how technical measurements transform when people start using and attributing meaning to new technologies. Charlotte Jensen

(2013, 2014), for instance, has shown how energy efficiency is not the only, or even the primary, concern when people light their homes. In a comparative study in Denmark, between people living in low-energy architecture, and another community focused more on social functions, environmental considerations and local management, Jensen showed how uses of electrical light may differ according to social values of sharing collective conventions in terms of energy and resources. Yet in both case studies the use of spot lighting technology resulted in more light sources, and thus did not necessarily lower the total household energy use. The important insight from this and related studies is that, while it is tempting to look for technological fixes when looking at energy consumption, it is also important to be clear about what is actually consumed. From a human-centred perspective, people may not so much consume *energy* as they consume artificial *light* (which consumes energy) (Bladh 2011; Bladh & Krantz 2008; Jensen 2014: 17). 'Energy is important, but it is what light facilitates and supports that make sense in an everyday life perspective' (Jensen 2013: 2269). This book shows that one such facilitating factor is the attunement of atmospheres.

The main purpose of this chapter is to outline a range of dominant approaches to light seen in social science and humanities, focusing on how light is entangled in social practices and meanings as an addition to the more technical, medical and aesthetic approaches. While daylight may be seen to be separate from human life, the way it is moulded in architecture, avoided by clothing, or prolonged by artificial lighting, are essentially socio-cultural and socio-technical processes and procedures. Light is entangled rather than separate from the social lives it constitutes. The chapter illustrates how light is not only shed *on* things – as in creating more energy-efficient light – but also *for* things, in processes where social norms and preferences change or adapt such technologies. The chapter therefore takes a closer look at the role of light in solving social problems, such as security in urban spaces, and the social life different technologies originate from or have helped forge. In essence, then, the chapter outlines how a social approach to light highlights central themes about living with light, both at home and in the city.

To do this, we move from the light-as-energy approach above towards a second dominating approach to lighting, which I define as 'light as health'. Here, light becomes part of a scientific project to understand the nature of light and how shedding light on human bodies influences human psychology and physiology. Such a perspective has also come under the scrutiny of social science for its consequences and reliance on ideologies of shaping social subjects. After exploring how social sciences have approached this second dominating perspective, a third focus is covered: safety and security. Here it becomes evident that lighting technologies are used to solve social problems, yet this entails dealing with inherently social issues of power. Energy, health and security are thus dominant themes in lighting research, and also within social science. Yet there are also other approaches and perspectives presented subsequently, including the many things to be learned about the role of light in social lives from historical accounts of transitions in lighting technology and ways of relating to light, which has been the major contribution from the humanities. From this historical perspective, we approach contemporary society to review some of the recent studies on the sociality of light and lighting. The chapter ends by highlighting how exploring light is not meant as an opposition to studying darkness but rather as an attempt to show how light and darkness unfold in a contextualized spectrum that both informs and emerges from social life. No single chapter can of course include all relevant literature, but I have aimed to highlight what I find to be the most illustrative examples of the themes highlighted above. First, though, the theme of health and lighting is explored.

Healthy light

Martin, a visually impaired interviewee, perhaps best illustrates the light-as-health approach. Martin, who has no sight or light sensibility at all, finds his way around the apartment without light. He simply does not need it. Yet he has a special lamp mimicking sunlight that automatically turns on during the afternoon and off in the late evening. He has a vitamin D deficiency, leaving him easily tired. He is not outdoors often, and, based on expert advice and extensive research on sunlight and health, he reasoned that he needs more light than he gets from the sun coming through the windows. This link between vitamin D and sunlight has long been known, and has led to light, daylight in particular, playing a central role in what has been termed healing architecture (cf. Carter 2007; Lawson 2002). The lamp thus offers him light on his body in his everyday indoor life, while he also reasons that it shows the surrounding neighbours that there is someone living in the apartment.

In Scandinavian countries such as Denmark the sun shines for six and a half hours on the shortest days of winter, resulting in long nights and the associated condition of seasonal affective disorder (Boyce 2014: 538–9; Tonello 2008). Knowledge of the importance of sunlight in Denmark – acquiring it during winter and avoiding it in the summer by using a sunscreen or shade – is widespread. There are regular campaigns directed towards disparate groups such as office workers, the visually impaired and those members of religious groups, particularly the Muslim community, who cover their skin, to remind them to get their fair amount of sunlight. Light in this sense is a medical concern and a material through which health and well-being can be moulded

(Boyce 2014: Ch. 14). This also entails a biological factor where for instance age plays a central role in how people perceive light, elders needing more and different kinds of light (Lewis 2015; Lewis & Torrington 2013). Yet there is often a discrepancy between the light-as-health and the light-as-energy approaches, when, for instance, more light is the solution for health issues, but which raises energy consumption, either by having to use more electrical light or by using more energy to cool down places from the heat of sunlight. Architects such as Carlo Volf suggest that 'instead of optimizing the daylight, it is about balancing the daylight, balancing light and darkness during the day and the night' (2013: 12).

The shift to new lighting technologies has also meant the rapid development of new ways of lighting private and public spaces. There are now Wi-Fi controlled lights, day rhythm lighting and wake-up light alarm clocks that start in the morning, mimicking the spectrum of light from a sunrise, offering a much smoother transition phase from sleep to wakefulness by promoting melatonin and serotonin production. Such technologies or architectural moulding of light are essentially concerned with optimizing human life to increase productivity, creativity and alertness (a fraction of this work is seen in Begemann et al. 1997; Boyce 2014; Huiberts et al. 2015; Smolders et al. 2012; Smolders & de Kort 2014; Steidle & Werth 2013). Switching light on to achieve such productivity has been argued to be reminiscent of biological hacking by switching people on (Dutson 2010: 46) and Western societies are designed in the light of such knowledge, even if people do not recognize it in everyday life.

Such relationships between well-being and light are also entangled in social negotiations. As a recent case of citizen protest in London showed, the proposed construction of a 62-storey skyscraper has been caught up in 'right-to-light' protests among the nearby inhabitants. What is interesting is that they are imposing the right-to-light rules devised in the 1920s by light surveyor Percy Waldram, who decided that 'property owners should have at least enough natural light to be able to read an article in *The Times* with only a one foot high candle in the room as illumination'.[3] Or in the hospital sector one may learn that highly cultural ideas about 'home-like' go hand in hand with workspace lighting standards both to treat psychiatric patients and as part of healing hospital architecture illustrating the point in this book that lighting is about seeing something in a particular way more than simply seeing. (Parrott 2005; Stidsen 2013; Stidsen et al. 2011; Volf 2013). Or with my informant Hanne, who lived with her family in a suburb of Copenhagen, who liked to light candles because they made her relax mentally and feel comfortable. She did know that the candles also emit polluting particles that may be harmful to her and her family's health, but prioritized the mental rather than physical health (Figure 2.1). This illustrates that the history of such connections

FIGURE 2.1 *The well-being of candlelight in the living room. Photo by the author.*

between health, well-being and light also rests on cultural ideas, practices and material objects. Importantly, this is by no means a fuzzy, relativistic suggestion; it is only to suggest that a social premise, such as the act of reading a newspaper indoors, and what constitutes 'homely', are enmeshed in the spreading of particular kinds of knowledge and standards.

This entanglement between lighting practices, norms and science is perhaps best illustrated by Lars-Henrik Schmidt and Jens Erik Kristensen in their insightful book *Lys, Luft og Renlighed. Den Moderne Socialhygiejnes Fødsel* (1986, *Light, Air and Cleanliness. Birth of Modern Social Hygiene*). It is essentially a Foucault-inspired analysis of light and air as means of disciplining the subject. They explore the history of ideas and the rise of a new sense of hygiene during the nineteenth and twentieth centuries in Denmark, not as a claim for the irrational fear of dust and dirt, but as a particular approach into the history of rationality. The study shows how techniques and strategies of purification are implemented, at first shaped as rational purification against bacteria, before being turned into a form that is both aesthetic and ethical during the twentieth century. Schmidt and Kristensen show how some of the practices people perform today are deeply rooted in history through government projects, nineteenth-century light-as-health science, and various social movements and economic incentives (cf. Shove 2003). This was done through the promotion and institutionalization of the role of the housewife in making homes through cleaning strategies. It was also accomplished through fresh air (illustrated, for instance, by increasing ceiling height in apartments) (Corrodi & Spechtenhauser 2008: 51) and access to light (illustrated by the development and use of sanatorium architecture) became central to the construction of ideals of cleanliness in the nineteenth and twentieth centuries (cf. Carter 2007). What emerged as lighting *on* domestic areas to eradicate bacteria developed into lighting *for* the home as lighting was turned into a moral and aesthetic validation of the sense of hygiene carried by the family – the housewife in particular – responsible for cleanliness both as a bacteriological and moral issue, as further discussed in Chapter 8.

Light as security

A third dominant theme has been light as a security measure, both psychological and infrastructural. In noting that light materializes power, Tim Edensor rightly states that lighting arrangements 'testify to a dispersed, mundane form of power through which meanings and sensations become unreflexive manifestations of common sense' (2017: 107). By inscribing inequalities across space, light is part of surveillance and policing and is mobilized to assert aesthetic judgements and forge normative arrangements. Street lighting has been instrumental to shaping order in the city as illustrated below, demanding ever more lighting. But there is also an aspect to such lighting where poorer areas of a city may have either little lighting, due to municipal budgets, or quite advanced lighting to curb anti-social behaviour. The very act of lighting a street may thus display the economic inequalities of urban space in a topography of wealth (Bressani 2015; Edensor 2017: Ch. 4), while darkness in cities is often where danger, poverty and lack of order rest (Edensor 2015a; Morris 2011; Shaw 2015; Thornton 2015). Light in this sense helps to shape what people attend to and what they ignore. By doing this light, and the energy network of gas pipes and electrical wiring it is embedded in, becomes a powerful political tool, and this has been the case since the earliest days of illuminating the city.

For instance, in his seminal work on lighting, Wolfgang Schivelbusch (1987, 1988) goes beyond the history of a technological approach (Bowers 1982; Dillon 2002; Multhauf 1985; Perkowitz 1996; Russell 1968; Thwing 1959) by exploring social problems such as violence and theft, which new lighting technologies helped to mitigate (and create), and their reception and responses among the general population from the medieval era onwards (Bouman 1987; Brox 2010; Edensor 2017; Isenstadt et al. 2015; Koslofsky 2002, 2006; Nye 1990). His work on street lighting (1987) stands as a major contribution to understanding how particular historical circumstances have come to inflict on the symbolic potency of light centuries later. Schivelbusch shows how political power is distributed through technologies by exploring the social origins and reception of a technology, rather than viewing technology as somehow detached from society. For example, he elucidates the changing role of light in European cities. In the medieval period, the city had a time termed *curfew* between sunset and sunrise, and, as the term implies today, largely resembled a lockdown. The gates were closed and the night watch walked the streets to secure order, arresting people who did not carry a light with them. Then, during the sixteenth century, things began to change. People still had to carry a light with them when moving around, but the source of the light became spatially fixed, since house owners were now responsible for illuminating their own premises; not unlike the current obligation in Denmark to clean the pavements from snow in the winter (1987: 62). By the end of the seventeenth century, administrative public lighting was introduced in Paris. But at this point it was not because of ideas about 'health', as such, even if in the following centuries it tapped into ideas about bacteria and illness (see also Bouman 1987: 13; Zardini 2005: 21). Rather it was the effect of the establishment of a police force to secure law and order in the growing cities. Light became a tool of exercising law and order, and in that sense, became associated with the repressive powers of the police and rulers. Schivelbusch even raises the suggestion that the symbolic aspects of light as power overlapped with the quantity of light, so *symbolic* light was seen as *actual* light (1988: 95), meaning that when light was seen as all present in Paris, it was because the control of the ruler was all present.

Unlike the street lights in London, extending from or near the houses, the street lamps in Paris hung in the middle of the street and came to represent the power of the Sun King, Louis the 14th. The link between power and light also meant that there was widespread destruction of lanterns, which in Paris, unlike other places, was punished by the galleys. While one could orient oneself in the street, it was unlike today; in their fixed positions, the lights were used to create orientation and identify people, while the mobile versions were still in use, and a torchbearer could be rented to follow one *en route*. By the eighteenth century the Parisian torchbearers had also become police informants, again illustrating the link between power and light. The close link between the ruling powers and street lighting in Paris was perhaps most potently illustrated in the early days of the French Revolution when the nobility were hanged from the lanterns, rather than from trees or street signs (Schivelbusch 1987: 65–6, 1988: 97–106). What is even more interesting is how this idea about security is also connected to ideas about modernity. To be a modern city is to be an illuminated city. Clearly visible from the oft-cited NASA Earth at Night maps, development and light go hand in hand with light and security, and all the symbolic and metaphorical power they entail (e.g. Blumenberg 1993).

It is not difficult to recognize the links between security and light, also beyond historical studies like Schivelbusch's. A first step was the conceptual separation of light from danger, which the flame had hitherto presented an amalgamation of. With electric light the risk of fire has so dramatically decreased compared to gas, oil, or candlelight, that it no longer makes sense to conceptually conflate the two. Now light can easily be turned on without concern. Cecilia, from the introduction, would, despite her blindness, turn on her light to feel secure against burglars, but would rarely light her tea lights. Light mimics physical presence, and thus creates a sense of security and awareness. Most anti-burglar advice incorporates suggestions to install movement sensors or timers that mimic normal lighting activities at home.

Illumination from the homes also adds to the sense of security on the streets. Casper Ebbensgaard has shown how light from people's homes 'help the street' as one of his informants notes (2017: 178). Whether drawn curtains provide illumination or not, the light from homes adds ambient light. This luminosity, Ebbensgaard argues, 'is not only a case of seeing better, making the streets "brighter", but making them feel safer' (2017: 178). As Ebbensgaard further argues, such a sense of security or safety has shaped public spaces in acts of beautification, gentrification or decriminalization, when socially derailed areas are lit (see 2015, 2017). As part of a defensive urban infrastructure, light is used to deter drug users or youths congregating and thereby create a sense of safety, or to shed light on dark areas of the city, where shady activities take place.

What is interesting, however, is that the supposed effect of more light leading to less crime is by no means straightforward. In other words, lighting does not entail that urban atmospheres and behaviour are passive outcomes of lighting design strategies, but rather that the staging is met by diverse segments using and challenging such designs (cf. Hasse 2012; Schivelbusch 1987). In London, Joanne Entwistle, Don Slater and Mona Sloane have shown how the idea that light may be shed in public spaces to prevent anti-social behaviour, and to make places feel 'safe', may actually have the opposite effect, where people become aware that it is a potentially dangerous place. Where dark streets in more affluent areas such as Pimlico are seen as a luxury, the bright light in social housing facilities that allows for better CCTV pictures, is seen to display the social inequalities of the area.[4] Criminal activities may even increase as an area becomes more lit, since it also means more people and movement in that area, even if people may *feel* more secure in a better lit environment. Further studies on the relationship between crime and light have suggested that while light directly aimed to prevent crime may under certain circumstances be successful, a lighting strategy that focused on social worlds coming into being was not only efficient after dark – as lighting addressing crime would be – but also prevented crime during the day. This suggests that 'a theory of street lighting focussing on its role in increasing community pride and informal social control may be more plausible than a theory focussing on increased surveillance and increased deterrence' (Welsh & Farrington 2002: iv). In essence, there is no doubt that particular kinds of light *on* surfaces such as skin may change behaviour. Yet it may also attract other activities, or simply shift the problem and direct it elsewhere, and maybe the starting point should not be light on the individual person.

To summarize, approaches such as light as energy, light as health and light as security may be viewed in different ways. First, there seems to be an ideology *of* light, where light is linked to ideas about health, modernity and even truth. Without light, there can be no healthy productive state or person. Thus, in the ideology of progress we need to shed light *on* the world. Second, there is the ideology *by* light. Here light becomes a tool to achieve ideological goals such as law and order. In other words, light enables ideologies to take form, by shaping power, order and discipline.

Towards a social comprehension of lighting

Light as energy, light as health and light as security represent the vast majority of scholarship on light, while social science and humanities research is particularly present in the historical accounts of technological changes, as witnessed above. The technical and scientific approach to light is predominantly about cause and effect – light leads to security, light leads to health, energy-saving lighting lowers energy consumption. In another strand of research, scholars of science and technology studies (STS) illustrate that no technology is external to society or to cultural processes. Rather, as Wiebe Bijker has shown, social negotiations, for instance between relevant social groups of engineers, designers and users, lead to compromises and different perspectives on what constitutes a problem, while eventually also influencing society by shaping official standards and social norms of visual comfort (Bijker 1992: 98, 1995; cf. Hughes 1985; Reich 1992). Technology and human culture are not separate entities, they are integral to one another, and market forces, political power and (bygone) regimes of knowledge have had a central influence on organizing the world as people experience it today.

Beyond the above approaches, and less known than the historical and STS studies of light, at least in an Anglophone context, is the seminal work on light and darkness by Swedish ethnologist Jan Garnert (e.g. 1993, 1994, 1997, 2016). The central point for Garnert in his *Anden i Lampan* (1993, *Genie in a Lamp*) is to show how lighting frames the context for human action and ideas, and how different lighting technologies take part in the human organization of time and space, particularly in the rural areas of Sweden. This ranges from organizing daily rhythms, to more seasonal rhythms of preparing for the winter during summer and repairing tools and clothes during the long winter nights. In many ways Garnert marks the beginning of a fuller understanding of an everyday lighting, beyond power, energy consumption and individual health and psychology. His historical study of texts and paintings from the eighteenth century to the 1990s indicates that it is not as such illumination that is the vital element of an ethnographic study, but the way light reconfigures time and space in everyday human lives. Garnert claims that light, particularly before the advent of electrical light, offered a 'compulsory togetherness' in Sweden (1997: 64; my translation of the German 'Zwangsgemeinshaft') since values of thrift led to people gathering around light. An empty space was a dark space. Besides space, light also structures time as natural light set certain limits in the pre-industrial era, which then tied in with human activity.

Garnert shows, for instance, how lighting technologies and sexuality are related in terms of birth statistics. He exemplifies this with birth statistics showing how eighteenth- and nineteenth-century children were most often conceived in the dark months of winter and born in September, except those children born outside of marriage (*oäkta* in Swedish, meaning illegitimate) who were conceived during the long summer days of work in the field. From 1776 to 1860 most children born outside of marriage were born in March. Garnert also illustrates this relationship between darkness, time and sexual activity with the electricity rationing in the first quarter of 1970 in Sweden leading to increased birth rates in November and December 1970 (1993: 108). Akin to Bijker's approach, where sociality and technology are interwoven, Garnert emphasizes the way in which technology, even such a seemingly intangible one as light, affects human behaviour and the way people organize their lives.

Garnert, Schivelbusch, Bijker – and many other researchers exploring the historical transition to electrical lighting (e.g. Barnaby 2016; Olesen & Thorndahl 2004) – show how lighting is not simply a technological tool to illuminate the world, but likewise a technology that has social consequences and origins. This was by no means only a Western phenomenon, as illustrated by Xiujie Wu's (2008, 2009) work on light in China. Her study of the introduction of electricity in a village in northern China in the 1970s ties light into political ideas about development. The introduction of electrical light offered to widen the visible domestic spaces in the evening. The quantity of light being shed in the houses rapidly increased, since the cost of electricity – as soon as the electrical infrastructure had been established – was cheaper than kerosene or oil. Yet the economical thrift of earlier times in terms of not wasting oil or kerosene, was transferred into electrical light quantity preferences. The new electrical lighting came with a 15W bulb, much more than previously even imaginable, but even though the younger population wanted brighter light it was seen as a waste to have a 60 watt bulb (Wu 2009: 148).

Yet even more interesting than the cost or quantity of light, was the behaviour it elicited. Women had previously controlled the oil lamps in order to do their housework and handicrafts, moving their lamps to places where tasks were performed. The man was largely part of the public sphere in the evening, performing his tasks elsewhere than in the domestic sphere. However, the new electrical lamps were fixed in particular spots, whereby women now had to come to the light, rather than, as before, taking the light with them. Wu argues that women experienced a loss of control over their lighting technology, leading to a blending of previously gendered spaces where the men and women would now remain together without disturbing each other's tasks. Moreover, as also touched upon in Chapter 7, the responsibility for dealing with electric appliances in light of the risk of electrocution was left to the men as a sign of 'the courageous self-sacrifice of the protector of a family rather than an exhibition of masculinity through a new technology' (Wu 2008: 220). Once again, new lighting technologies did not simply shed more light on the world but also contributed to changing social roles and lives. Yet they did not do so in deterministic ways, as there were also cultural preferences that resisted the mindless adaptation of new technologies, such as thrift, nostalgia embodied in everyday practices, and ideas about gender. In that sense, the history of technology also becomes cultural history.

Textures of luminosity

What is so intriguing about light – besides its very quality and quantity – is the way light in all its facets of colour, shadow, luminance and glow, interacts with surfaces to shape human perceptions of the material world surrounding them. Junichiro Tanizaki in his seminal book, *In Praise of Shadows* (2001), originally published in 1933, reflects upon the transition to electrical light, lamenting how the traditional Japanese appreciation of shadows, glow and luminance in material objects was replaced by a Westernized intense illumination. He writes about how electrical lights replaced the dimmed lighting from candlelight for which the restaurant was famous, even if the electrical light was put in old-fashioned lanterns. Insisting on experiencing the famed, gloomy illumination, he used candles because 'only in dim half-light is the true beauty of Japanese lacquerware revealed' (2001: 22). Hence, he realized that speaking of light is also to speak of the power of darkness, glow, absorption and reflection – in essence, luminance as an indispensable element of a material surface and its appreciation. His critique is not of the convenience that electrical applications result in, as such, but rather, in his somewhat nostalgic writing, that the design of lamps and light levels did not consider cultural habits and tastes (2001: 12). In this, Tanizaki reminds us that surfaces of design objects are made to interact with light in a certain way. In the wrong light, a wooden surface may appear lifeless and cold, our skin bleak, a coloured dress appears a different colour, and many of our choices and actions are done in a light that suits that task and presents the surfaces of material bodies in the culturally preferred manner.

Material surfaces and colours reflect light differently. Nicholas Saunders (2002, 2001), for instance, has shown how the role of shimmer and iridescence in materials such as obsidian in pre-Columbian Mesoamerica played a role in bridging symbolical and physical realities (cf. Kapstein 2004b; Kumar 2015) and linking material properties to landscapes, deities, myths and everyday life, while also changing roles and meanings as political changes occurred. As a consequence, while people see light as it is reflecting on/off surfaces, it is likewise clear that understanding light also means understanding core features of materiality: the texture and colour of surfaces, including the ideas of visual presentation that the designers had in mind; that is, the kind of light the objects are entangled in, when in use.

The above historical and ethnological studies thereby reveal insights into the social history of light and lighting, showing how gender, power, symbolism, religion and everyday practices are entangled in technologies, both in their construction phases and in the subsequent use and reconfiguration. Yet gazing historically on light, whether in terms of health, energy, security or other themes, also leaves certain aspects inaccessible. As Claire Nesbitt notes in terms of studying light in a Byzantine church as an archaeologist,

> Light is difficult to study for it is manifest only in its impact on other matter. Light leaves no evidence in the archaeological record and to understand the traces it leaves in Byzantine literature is to rely on interpretation of what often seems to be subjective descriptions.
>
> 2012: 140

Historical studies are equally able to reveal the changes and larger cross-cultural issues of the introduction of new lighting technologies that bring important new insights into 'what good lighting is' and has been, along with the social lives it has taken part in shaping. These studies show that what is considered proper lighting – that is, lighting as it takes part in constituting aesthetic appreciation and social activities – is socially informed.

This is perhaps most eloquently shown in terms of the different lighting standards across the globe (see special issue of *Energy* 1993, vol. 18, 2). For instance, lighting standards for many tasks have been twice as high in the US compared with e.g. Norway (Bartlett 1993: 173). Mithra Moezzi has also shown how recommended lighting levels for certain tasks in the US have increased radically: 'for classroom chalkboards the 1938 recommendation was 150 lux, whereas in 1972 it was 1400 lux; the 1930 recommendation for fine detail drafting was 175 lux, whereas the 1970 recommendation was 2000 lux' (1998: 8128). But the interesting thing is not so much that light levels have increased (this may be the result of more research on biological responses, or the ability and finances to prioritize the installation of more light). The interesting thing is how historical, and thereby social contexts, take part in defining them. Moezzi notes that since 1970 the US recommendations for lighting levels have decreased again by as much as 50 per cent (1998: 8128). As mentioned earlier with the oil crisis in the 1970s, new building regulations in Denmark demanded smaller windows being used to reduce heat loss, leaving more darker spaces with less natural light (Kristensen et al. 2004: 115). Or when words such as 'home-like' (*hjemlig* in Danish) are part of the official Danish lighting regulation in hospitals (DS703) (Stidsen 2013: 20). These more socially informed perspectives on light illustrate that it has been different in the past, and that social worlds both help to shape, and are shaped by, technology (even though they may be seen as exempt from such social origins as purely technological inventions). The historical studies make one rethink the way contemporary societies are dealing with new lighting technologies, old ones, the social subjects they create, the meaning they have, and the visual presences of the world that both artificial and natural light produce (cf. Nye 2018).

Light cultures beyond energy, health and security

Contemporary studies have discussed the multivocality and social life that luminosity shapes across the globe (Bille 2015b, 2017; Bille & Sørensen 2007; Gage 1995; Isenstadt et al. 2015; Kumar 2015; Pink & Mackley 2016; Shaw 2015; Young 2006). The heightened focus on energy consumption in social science has proven its relevance by exploring people's domestic practices rather than producing technical calculations of potential use with the premise of 'all things being equal'. Such a premise is rarely the case in real life. As Kirsten Gram-Hanssen found, something happens to low-energy buildings when people move into them, and the energy demand does not follow the higher energy efficiency (Gram-Hanssen 2008a, 2010, 2011, 2015).

Perhaps it is no surprise, then, that both the presence and the absence of light carry potent social meanings and possibilities as places for gathering; daytime affords certain practices, night-time others (cf. Chappatte 2014; Edensor 2013a, 2013b, 2015a; Galinier et al. 2010; Shaw 2013, 2015; Winther

2008). Yet it is unwise to separate lighting practices in urban (Western) cities from those in rural areas in India, the Middle East, or in Africa. These places may bring attention to issues overlooked in the West. Not because these issues are not extant – illustrated by the new modes of lighting offered as immigrants move into residential areas (see also Chapter 6) – but because most people are so used to lighting that the analytical gaze is opened only as difference is occurring. This, I find, is the task of a social approach to lighting: exploring the meanings, effects, practices and historical shifts unfolding as people engage with, or through, light, in deliberate or routinized ways.

With a particular eye for the most taken-for-granted meanings and practices in everyday life, social science approaches to contemporary lighting have highlighted (at least) four features of a socially embedded light perspective: 1) since light offers visible spaces for cultural transmission, changes in lighting technology also restructure the social organization of time and space; 2) emotions and cultural notions of well-being are central to lighting practices; 3) visual comfort and lighting are entangled in the very practices and routines of staging illumination; and 4) issues of class and identity unfold through lighting. I will turn to each of these below.

First, an important step towards a more social approach to lighting is to abandon any binary distinction between light and darkness. It may work in metaphoric and symbolic uses of light, but by looking at the way people orchestrate and experience places, there is no clear distinction. Total illumination makes it impossible to see the world due to blinding effects and lack of shades, in the same way as total darkness makes nothing visible. When I talk about light, I thus by definition also talk about elements of darkness, shades and textures, of both artificial and natural lighting. Most often people are surrounded by just a little bit of light, often nothing more than just a reflection from a surface, sky or star, and everywhere there is a spot in shade or darkness. What matters is the balance and dispersion of these landscapes of light, what I have called lightscapes (2007). Lightscapes offer a lens through which to understand the practices of visually designing spaces, informing social lives, while light from the perspective of the lived experience of places is integral to perception and the very identity of a place (cf. Ingold 2011: 134, 249). Once again, this goes to show that people often do not switch on lights, draw the curtains, or design a house with fenestrations just in order to see, but also to attune the space they are occupying, to promote particular actions and visions.

Wiessner (2014) follows this line of thought and shows how particular social practices and intimate conversations are handled at night among the Ju/'hoansi Bushmen in South Africa. What could appear to be an economically unproductive fire lighting practice in the evening turns out to be socially productive: it helps transmit cosmologies and nocturnal rules of behaviour, while daytime conversations are centred on economic topics and gossip. Tanja Winther (2008) similarly shows how the introduction of electrical lighting in rural Zanzibar helped combat an increasing depopulation and rapidly changed the social spaces, both public and private, in light of the new nocturnal sensorium. Here, cultural notions of safety, honour and gender take centre stage (cf. Kumar 2015). Conversely, illuminating the night also becomes a way of expressing hierarchies, where the illumination of a school, a mosque, a former politician's house and a successful tourist guide's residence comes to signal the social status of these people and institutions. Although there is a shared wish to develop the rural town of Uroa in Zanzibar, what that development should involve, exactly, remains contested. Lighting is clearly an element, but light also 'represents social power and those left in the shadows can do little but watch how differences increase' (2008: 146).

Second, the above studies often focus on the material infrastructure of the home since the home is also where ideas about identity, norms and morality are forged (Blunt & Dowling 2006; cf. Cieraad 1999; Miller 2001). The ability to reveal and conceal the interior of the home is orchestrated by light and visibility, most notably through the windows (cf. Garvey 2005; Vera 1989). The home, more than the house, is an emotional relationship established through continuous practices. As will be clear throughout this book, people are continuously attuning themselves to their home, buying new things through which they (re)present their individuality, family and group affiliation. Common practices, such as getting ready to watch television, making coffee or dinner, are all part of attuning oneself and the space to the activities that are, or should be, occurring in a particular way (Bille 2015b; McCracken 2005; Olesen 2010; Pennartz 1986).

Cross-cultural studies also show how lighting accommodates ideas about well-being, honour and privacy in markedly different ways (for instance in Denmark, Bille 2015b, in Jordan, 2017; in Japan, Daniels 2015; in India, Kumar 2015; or cross-cultural comparison in Wilhite et al. 1996). In her study of the Dayak longhouse in Borneo, Helliwell (1996) shows how light (and sound) form a central part of understanding people's social well-being. Absence of light from a house in the village stirred communal anxiety as to whether the person was ill, to the point of light being indexical of well-being. In this sense, the absence of light is not the same as darkness, as the absence points to the expectation of light, rather than to degrees of darkness, which have been a condition of night-time throughout most of human history.

This focus on the role of light by no means indicates a lack of attention to darkness or shadows – quite the opposite. Darkness allows for some things to happen rather than others, precisely owing to the subdued or lack of lighting (Bradley 1989; Cook & Edensor 2014; Edensor 2013a, 2015a; Edensor & Falconer 2015; Ekrich 2006; Galinier et al. 2010; Koslofsky 2006; Morris

2011; Shaw 2015, 2018; Williams 2008). Edensor (2017) in particular has explored darkness in his studies ranging from dining in the dark at a concept restaurant in London, driving and cycling at night, to walking the dark forest of the Galloway Dark Sky park. Here, the particularity of being in usually luminous landscapes is central for events to unfold. He illustrates how the senses of a modern urban population are reconfigured, and how people essentially experience the world differently in an environment where there is unusually little light, and such reconfigured experience has been forged in the context of living in the city. Edensor notes how in a 'sensual and aesthetic reconfiguring of the night, darkness is complementary to light, a sensual (dis)ordering and revaluing that contrasts with the rationalising and moral bourgeois order imposed by illumination' (2017: 177).

Notable is also Wilhite et al.'s (1996) seminal work comparing Norway and Japan to showcase how different energy practices rely on cultural ideals, from the focus on lighting and multi-room heating in Norway to create cosiness, to washing and dishwashing in Japan. Wilhite et al. show how, on average, the Norwegians have 9.6 light sources in the living room compared with the 2.5 lamps in Japan. The Norwegians often noted how a ceiling lamp was 'too cold', and, much like in the Danish homes, created rooms within the room through light. The wish for a particular cosy atmosphere in Norway, termed *koslighet*, resembling Danish *hygge*, became 'a cultural energy service', which Wilhite et al. define as 'a set of energy use behaviours deeply rooted in the social, cultural and symbolic presentation of the home' (1996: 798). The incandescent light bulbs were particularly preferred in Norway for offering 'warm light', thus having the ability to make a cosy atmosphere. That said, more people had fluorescent light in the toilet and kitchen, because it offered more illumination to see things. In Japan, it was the opposite. The incandescent light was seen as 'cold', offering an atmosphere associated with hotels because it was too dark, and made it more difficult to see the normal features of the surroundings, where the informants 'like to see what I am eating' (1996: 800). Thus, the Japanese used bright fluorescent light in places where social interaction took place and incandescent bulbs in other places, such as the toilet and kitchen.

Third, while studies of the role of light in rural areas of non-Western countries may help inform us on how light is deeply embedded in social life, there are also plenty of examples within Western society that lead us to question taken-for-granted concepts and scientific knowledge. One such case is found in Philip Vannini and Jonathan Taggart's study of off-grid households in Canada (2013), which illustrate how ideas about comfort are not simply about visual responses to a lightscape but are similarly entangled in degrees of participation in the production of the electricity that goes to generate the light. It may be that, from an 'all things being equal' perspective, one may reduce human perception of light to visual parameters; however, what Vannini and

Taggart also show is that the materials, skills and meanings associated with that light are paramount. From this it emerges that light intensities must be seen in concert with the practices that came to produce them. Following Crowley (2001) and Shove (2003: 57–61) it becomes clear that comfort is malleable, and it is only recently that it has been normalized, particularly in the hands of the producers and technical experts 'perceived as natural and fixed, as if an optimal standard could be positively measured and assessed' (Vannini & Taggart 2013: 1082). To illustrate the point of linking practices with the perception of comfort, they note that 'the key to achieving various intensities and different kinds of off-grid visual comfort resides in off-gridders' capacity to reassemble the sociotechnical system by which comfort is achieved within their domestic life' (Vannini & Taggart 2013: 1080).

Central to understanding the diverse cultures of light is also the role of routines – the everyday practices of getting up, lighting one lamp, going to the toilet, lighting there, pulling the curtains, opening windows, getting dressed, etc. (Gram-Hanssen 2008a; B Hauge 2013a; Hauge 2015; Shove 2003). Sarah Pink has in this context advanced our understanding of the home as a sensuous encounter that forges social roles and identities (2004). Pink and her colleagues have, in recent studies, explored the night-time routines of people going to bed; what light they turn off, when, and why (Pink et al. 2015; Pink & Mackley 2016). The studies illustrate how *care* for other people is a key part of people's practices with light; borrowing light from one room to the next to make as little disturbance as possible, and bodily tactics to remember all the tasks take the form of routinized movements through the house. These studies thereby point towards movement *through* the home as part of routine practices at home beyond the more common approach of what people do *in* the home. They also illustrate the sensuous work being done at the margins of attention. Such studies bring attention to the roles of light and darkness in notions of security, comfort and well-being, and how they should – at least from a human perspective – be seen as a relational practice, rather than as a passive objective resource that the body simply registers (see also Gram-Hanssen 2008b, 2011).

Finally, Edensor's extensive work on lighting shows how issues of class and identity are central features of space-making processes through light (Edensor 2012, 2013a, 2017, Edensor & Millington 2009, 2012). Many of his studies revolve around the role of light in perceptions of landscape and space, but other studies deal directly with how light takes part in performing and confirming ideas about class. Aside from the study of extravagant Christmas lighting explored in Chapter 7, Edensor and Millington (2012a; 2012) visited the Blackpool Illuminations, an autumnal festival with massive light displays along six miles of the seafront, visited by more than three million people every year. Participation in such a light display becomes closely associated with taste, nostalgia and family life, illustrating how ways of seeing and visual preferences are entangled in social validations. In essence, then, particularly in the work of Edensor, the importance of understanding the social life of light is pivotal, as it shows how light is not simply neutral – neither in use nor in display – but tied up in ideas about subjectivity, time and space.

Conclusion

The chapter has illustrated how particular kinds and intensity of light – fixed or movable – allow for diverse meanings, practices, atmospheres and social lives to unfold (Bille & Sørensen 2018; Kapstein 2004; Pink & Mackley 2016). Humanities and social science studies like those reviewed above offer a tool for understanding people's lives, their life worlds and the central role of materiality in shaping cultural spaces. In a poetic attempt to understand different luminous effects, it has recently been suggested that,

> [d]arkness tells us something about ourselves. What we essentially see when it is pitch dark is our own hallucinations [. . .] The shadows, on the other hand, are a tremendously efficient way to tell us something about the world [. . .] shadows are a highway to see the outer world. Darkness is a highway to see the inner world.
>
> NØRRETRANDERS & ELIASSON 2015: 25, my translation

Light and darkness are embedded in social life; they influence, shape and normalize human practices and visual occurrences. What is needed in the contemporary debates about the role of light in human life, and 'what good lighting is' (most often centred on the psychological, biological and technical aspects of light), is a keen eye for the cultural differences, meanings and practices in which light and darkness are entangled. This is because the way people adopt or reject light may not always be in accordance with scientific claims of optimal light; not even if people know what is optimal based on those premises. Light is more than energy; it helps social life make sense. The argument I made with Tim Flohr Sørensen (2007) a decade ago was to look at the effects and embeddedness of light in the sensuous lives of people, the atmospheres it helps to shape, and the politics it is part of. This chapter has continued this call to see light in qualitative terms, by reviewing some of the literature offering social perspectives on light. As Richard Peters notes, 'we must persuade the public that the *quality* of light is far more important than the *quantity* of light' (1983: 43). Yet it is vital to understand that 'quality' can only become meaningful if the social, material and cultural practices and ideals are incorporated.

The chapter has illustrated how light is about central themes in human life, often at the same time: science, philosophy, language, art, natural history, culture, theology (Dutson 2010: 13). The important point to draw from this is the role of *people* in architecture, energy use and design, rather than reducing them to 'bodies that respond only to biological stimuli' (Dutson 2010: 67). As a result, the following chapter will turn to the light that shapes atmospheres and engulfs cultural terminology.

3

In the vagueness of *hygge*

Introduction

Susie, the bartender, stepped towards the table in Café Langebro in early afternoon on Islands Brygge with a lit tea light and proclaimed: 'Here comes the *hygge*.'[1] Along the walls are spread historical photos and signs from the past showcasing the development of the area as an industrial harbourside over the twentieth century, establishing a sort of authentic traditional working class pub oozing with an informal atmosphere of *hygge* (cosiness). This day the place was quickly filled as a power cut had closed down the nearby offices and workplaces. Since it was broad daylight in the café, it was clear that lighting the tea light was not about creating visibility, but about instantiating a particular way of seeing and feeling space through an atmosphere characterized as *hygge*. Susie's comment announced that a cosy atmosphere had arrived in the shape of a tea light.

If researchers were to take the words of their informants at face value, the tea light does not *represent* cosiness but *is* cosiness. Susie is not alone in Denmark in her conflation of concept (cosiness) and matter (light) – or the lack of separation to begin with. In lamp commercials in Denmark, one sometimes sees expressions such as 'turn on the *hygge*' (*tænd for hyggen*) and, in these days of economic crisis, one can buy '12 months of interest-free *hygge*' (*12 måneders rentefri hygge*) in the shape of a wood-burning stove with a glass front. Likewise, in home decorating magazines' coverage of Danish homes, dimmed, reddish light and cosiness seem more often fused than separated. From an emic (i.e. informant's) perspective, you can argue that Susie literally came to the table with *hygge*. Clear statements like Susie's soothe the researcher, leaving stability and fixed points in what is otherwise a chaotic academic process of figuring out life worlds from the generated data of what informants say and do.

There has recently been a rise in social science studies discussing the multiple natures of objects in human lives; not just different perspectives on the same objects, but how the objects may be seen as altogether different, yet related: a so-called ontological turn (cf. Henare et al. 2007a; Holbraad & Pedersen 2016; Law 2011; Law & Singleton 2005; Mol 2002; Viveiros de Castro 1998). To apply such a perspective to the above, consider the tea light not as *one* thing, which different cultural horizons have different perspectives on, but rather as an altogether different thing construed through the way people act with and talk about it. Researchers need, it is argued, with 'purposeful naïveté [. . .] to take "things" encountered in the field as they present themselves' (Henare et al. 2007b: 2). As scholars, from our etic position, we may analyse a certain material substance as a *symbol* of, for instance, cosiness, but from the perspective of the so-called ontological turn, researchers also need to take the emic perspective more seriously if informants construct a type of light, not as symbol of, but actually *as*, cosiness. Stating that cosiness is, rather than represents, light, may appear at first hand quite an intellectual challenge. But how else can one understand Susie's statement if not through a purposeful naïveté? In this way, it is argued, researchers do better justice to the intricate categorizations, practices and worldviews that the informants have and therefore may challenge common academic understandings (Henare et al. 2007b). To understand informants' points of view, researchers need to take what they say and do as statements about the nature of the world. By taking the words of my informants – Susie, the PR bureaux, journalists and other people – at face value, *hygge* and certain kinds of light become one and the same.

This chapter offers a conceptual exploration of the nature of *hygge* as a central part of domestic and social life in Denmark. But it needs to be stressed that the point is broader than simply to offer context and clarification of *hygge*. Rather it is to encourage to scrutinize the atmospheres people live through in detail: what situations do they classify? How are they assembled? What are the historical roots? Precisely because sensory and atmospheric experiences are hard to verbalize, how vague or precise is the definition then and with what social effects? Aside from clarification of a central, though not sole, atmosphere in Danish homes, the chapter takes Susie's clear expression that *hygge* and the tea light are one and the same as a starting point for a methodological exploration. We also look at what can be gained by scrutinizing expressions of clarity with juxtaposing expressions of ambiguity, vagueness and multiplicity – the 'I think *maybe*' or 'I don't know, *perhaps*' one often hears from informants, particularly when addressing issues of explaining sensory experiences and atmospheric engagements.

Take, for instance, another informant who while sitting in a home where much attention had been given to the interior design, noted that 'I have a hard time articulating *hygge*, but, really, it is an atmosphere. It is a feeling, which probably might be really hard to define in words.' This informant teaches us that *hygge* is an atmosphere and feeling but also vague and beyond words. As Andreas Rauh notes, 'it is difficult to talk about atmosphere when it comes to scientific descriptions. The atmosphere is not an object of research that is particularly objective. It is merely a hard-to-grasp phenomenon that requires all of a researcher's sensual awareness' (2014: 247).

Often such uncertainty, or lack of ability to explain oneself about what defies words, may be brushed away in publications, focusing instead on clear expressions or observations. This is a result of what sociologist John Law calls *definiteness* (Law 2004; Sørensen 2012, 2015), in which the sciences – social sciences included – often view lack of clarity as a methodological problem on the scholar's side. But as Law notes, what if the world is messy, consisting of vague objects and concepts rather than definite ones? What if informants are not quite sure what a thing or the 'world' in all its totality really is? What if a thing is perhaps more than just one thing for a single individual and instead multiple things at the same time? Or what if informants feel that the words they use – and fieldworkers record or jot down – are not sufficiently encompassing enough to describe their worlds, but nonetheless make do?

Exploring the role of vagueness in central terms, such as *hygge*, I argue that in understanding the informants' worlds we need to take informants' vagueness and ambiguity at face value. This still entails being methodologically systematic and thorough, but it highlights the analytical and theoretical potentials in the process of coming to terms with the messiness of informants' realities, rather than reducing or glossing over it. As a cultural category of an atmosphere, *hygge* is an exemplary case for exploring such juxtapositions between clarity and vagueness, because the ontological nature of atmosphere is also vague, and yet felt as part of reality. For instance, Marianne, an informant in her early fifties, noted that a homely atmosphere is 'that feeling you get the second you enter your home. It needs to be a comfortable place to be. You should *feel* at home. It is difficult to articulate. But it is a feeling'. Taking vagueness seriously should not mean ending with an inconclusive analysis. What is needed are precise descriptions of the way such vagueness informs social life, grounded in empirical data, as explored in the following chapters. This chapter argues that clarity and vagueness are not different perspectives on the world, but part of a 'contemporaneity' of meaning, presence and practice that is more fruitfully explored with a focus on what makes up the elements of atmospheres. By not being defined and prescriptive, but vague, *hygge* allows for a wide diversity of people to co-produce atmospheres in their own way. This is the social power of vagueness.

This chapter outlines the concepts of *hygge* as an atmospheric plurality encompassed in one term, and highlights the central roles of light, time and vagueness within it. Chapter 4 goes into more conceptual discussion about atmospheres as the bodily-felt emergence of attuned people and the spatially extended affective presences of matter. These conceptual clarifications of what are otherwise ontologically vague experiences and concepts, are needed in the remainder of the book to demonstrate how atmospheres – one specified as *hygge* in particular – are part and parcel of understanding the role of light in social life in Denmark.

Hygge as process

Hygge is a very common term in Denmark, and is central to Danish home design and social interaction. It is a concept that avoids strict definition, even if there is a core essence of referring to a relaxed atmosphere. Closest to a definition is perhaps Judith Friedman Hansen's description of *hygge* as,

> comfort, cosiness, cheerfulness, and friendliness. To be in a situation characterized by hygge is to be in a state of pleasant well-being and security, with a relaxed frame of mind and open enjoyment of the immediate situation in all its small pleasures.
>
> 1976: 54

It is thus a term used for a characteristic kind of informal and relaxed atmosphere. When not pursued in solitude, it is something that takes a collaborative effort to achieve, and implies an ideal of absence of conflict or open disagreement. In that sense, it may be seen as both erupting in, and enforcing an, egalitarian space, where threats to such absent hierarchies are often met with 'strong negative sanctions in the form of humor' (Hansen 1980: 167).

Despite being a very common term in Danish culture, comparatively little dedicated research has been done on *hygge* (Bille 2013a, 2013b, 2015b; Bille & Sørensen 2007; Hansen 1976; Levisen 2012: Ch.3; Linnet 2011). Carsten Levisen (2012: Ch. 3) has eloquently shown how Danes see *hygge* as a motivating factor for doing things, illustrated by the common way of doing things not necessarily because they are wanted, but for *hygge*'s sake (*for hyggens skyld*). For instance, despite being in a hurry to prepare for a dinner party, a middle-aged couple living on Islands Brygge would do shopping and cleaning together for the sake of *hygge*, instead of a more time-efficient delegation of the tasks. *Hygge* cannot be fully controlled, and even if light holds a central role as shown by Susie, it does not, by definition, cause *hygge* but is rather part of its emergence. As Levisen rightly notes, '*Hygge* is "fragile" because the process, in a sense, *is* the goal' (2012: 104). *Hygge* is thus a term every Dane knows and to which foreigners are quickly introduced, but it nonetheless escapes a single definition. My informant Henrietta, for instance, noted that: 'I think it is very difficult to define *hygge*. No wonder foreigners do not understand the Danes when we say we are about to *hygge*.'

It has become particularly glorified in a recent spate of international literature (Edberg 2016; Russell 2016; for a critial review of such literature see Sangild 2016). The tourist organization VisitDenmark has even established a *hygge*-council (*hyggeråd*) in 2018 with scholars and prominent people who will work towards an application for the UNESCO Representative List of Intangible Cultural Heritage of Humanity to include *hygge*. On a more national scale, *hygge* was included on the Ministry of Culture's Denmark Canon as one out of ten important values that has shaped Danish society, along with welfare society, gender equality, trust and freedom. Here it is stated that: 'Hygge is seen as a special way of being together in a relaxing, nice atmosphere.

Hygge has its own word and many people say that it cannot be translated.'[2] Scholars, however, often argue that the national exclusiveness is incorrect, since most notably the Norwegian *koselig*, but also the German *Gemütlichkeit*, the Dutch *gezelligheid* and the Russian *ujut* are closely related (Borish 1991: 276; Gullestad 1992: 79–81; Levisen 2012: 110; Linnet 2011: 35). The English word 'cosiness' does not fully encompass the feeling. While the term has clear national connotations, the historical roots of *hygge* are actually linked to a late-nineteenth-century loan word, derived from a Norwegian verb meaning 'to console, to encourage' (Levisen 2012: 80).

The commonness and centrality to social life of the term are also reflected in speech formulae such as the welcome or departure greeting 'hyggeligt at se dig' (directly translated into *cosy to see you*) or 'kan du hygge dig' (*can you cosy yourself*, equal to *have fun*) when departing (Levisen 2012: 87–90). It is mostly a term characterizing an atmosphere, as already shown, yet it is also extensively used in various grammatical forms to describe ways and goals of doing things, or the affective qualities of individual objects used to cover and validate aspects of social life. In essence, it is a cultural-specific concept through which socio-material life is organized (Levinsen 2012: Ch. 3).

Although *hygge* is a vague concept, it is mostly associated with positive things. But that does not mean that the effects of orchestrating *hygge* are only positive. As stated in a contribution to debate in the Danish newspaper *Jyllands Posten*, it is 'an insistent, soft broom that sweeps uncomfortable truths and disputes into the eternal darkness under the heaviest rug in the living room'.[3] *Hygge* may also be a contested and slightly condescending expression when it, and its implied relaxation, describe a too informal behaviour or passivity when action and formality are actually needed. In some contexts *hygge* is used 'to the point of being close to coercion' (Jenkins 2012: 41). Despite claims to the opposite – of relaxation and informality – it may be highly structured and planned, underscoring the seriousness of the effort to orchestrate and achieve it (Ardener 1992: 26–7). Susie made the gathering at the café *hyggelig*, but that is also the way it *should* be.

To offer some ethnographic context, Morten is a young student living in a somewhat scarcely furnished rented apartment. Despite the lack of economic position to fully materialize his ideals for a nice home he did prioritize designer lamps and furniture and an explicit focus on what may be termed the materiality of hygge with candles, blankets, and offering tea to orchestrate a relaxed interview situation. While he was quite aware of his use of things to orchestrate a variety of atmospheres, from emphasis on intimacy, to dinner party, or work situations, he still also had problems with *hygge*. He particularly found Christmas, which otherwise epitomizes a time of *hygge*, sickening when it appears as 'forced *hygge*'. As he expressed it, 'now everyone has to get their gifts and we *have* to get the best food while everyone is actually stressing and not enjoying it'. The problem is not that the atmosphere is staged, he does this explicitly himself as well. It is rather the excessive focus to orchestrate lack of conflict and what people should feel, rather than letting it evolve within a material setting. Excessive attention to perfecting a conflict-free atmosphere creates conflict.

Together with Levinsen and Hansen, Jeppe Linnet (2011) has offered the most extensive and nuanced discussion of *hygge*. He describes the historical roots of *hygge* as expressing middle-class values in Denmark, based on his interviews with four middle-class families in the suburbs of greater Copenhagen, and archival work. His research highlights how *hygge* rests on withdrawing social interaction from economic spheres and alienating modernist conditions, even if it is also part of structuring social control and distinction. *Hygge* is inclusive of those within its sphere, but it is also excluding those who do not know the codes. Linnet thus rightly points out that there are plenty of judgement calls based on social class about the ways different segments perform *hygge*. In such performance of class material culture are thus central elements in performance of class identity. As also stated on the Denmark Canon, 'it does not necessarily demand a lot of things to create a *hyggelig* atmosphere, for instance just a couple of candlelight or a bowl of Friday candy'.[4] So the question is not *whether* it occurs within different social segments, but *how*.

An example of the materiality of *hygge* may be in place here. Christian and Hanna are part of a middle-class family with toddlers living in a century-old apartment on Islands Brygge and working as public clerks. Their home is practically furnished in the sense that there needs to be space for all the toys, clothes, kitchenware, drying racks, books and pictures and so on. Yet they emphasize that it should also be a cosy space, that is, personal, almost cave-like, but not reminiscent of the minimalistic interior design otherwise popular in the neighbourhood. To them a *hygge* atmosphere is about expressing personality in a relaxed manner. This is achieved through selection of furniture, preferably with patina, rather than simply buying it in IKEA, even if they do have IKEA furniture. It is about establishing rooms within the room through lighting and furniture. Rather than something you just buy, it is a material practice where emphasis is on offering special tea or coffee, and on weekend evenings alone with the family to allow oneself to eat sweets and drink alcohol, which may otherwise be restricted. There is thus a materiality to relaxation, even if the atmosphere is not determined by such props. The whole process of getting attuned, preparing food together after a week's hectic work, and just relaxing with a class of wine and candy for the children while watching television, is exemplary of *hygge* to them. Nothing posh, just orchestrating relaxation. This type of family *hygge* is different from the hygge staged if friends would come by as this would cause more preparation and stress, when it entails orchestrating an atmosphere through offering all the right things. *Hygge* is about absences as much as about presences: an absence of worry, absence of conflict, absence of a market sphere – although it is simultaneously heavily embedded in consumption practices. The consumption aspect just differs depending on whether it is *hygge* with family, friends, at Christmas or alone. The material infrastructure and intensity of the atmosphere differ but the term is the same. *Hygge*, in this sense, is a classification of a variety of atmospheres where informality and relaxation is in focus.

In the light of *hygge*

A wide range of things and consumption practices are all part of *hygge* (beer, coffee, sweets, music, television, heat, etc.) but I focus here on light, as it plays a prominent role. To illustrate, the Danish tourist agency VisitDenmark describes *hygge* as follows:

> Hygge is lighting a welcoming candle when visitors arrive or meeting friends or colleagues in a candlelit café. It is cuddling up on the couch on a winter's eve wrapped in a woollen blanket with a good book and a mug of hot chocolate, or lighting up the stove and spending the evening with the family watching a good movie and eating sweets. It is the even, soft light of the PH lamps, designed not to shine in your eye, and the subtle light that Le Klint lampshades bring to the otherwise darkest corners of your home. Hygge is the smell of home baked cookies in the oven and wrapping Christmas trees on the front lawn with seasonal decorative lights. Hygge is about bringing light to the darkness and warmth to the cold.[5]

As evident from such tourism promotions, lighting plays a key role in *hygge* and social life in Denmark, particularly the abundant use of candlelight putting Denmark at the top of the candle-consuming countries, with an average use

FIGURE 3.1 *Photo of cosy lighting. Photo by Matilde Lykke.*

of 5.79 kilos per capita, significantly higher compared to the second country on that list, Austria, with 3.16 kilos.[6] At Christmas time the average use of candlelight triples. In autumn and winter, a recent survey showed that 28 per cent of Danes light candles every day, with 31 per cent lighting more than five candles at a time.[7] Anne Bay, director of the Danish Lighting Centre, notes:

> We really like to *hygge* [verb] in Denmark. The concept is a large part of our self-understanding, and we use the word to describe more or less every pleasant and relaxed atmosphere or event. Actually, we are so fond of *hyggen* [noun: *the* cosiness] that it affects the way we light our homes. We are really good at creating a *hyggeligt* [adjective] light.[8]

Aside from illustrating how *hygge* appears in various grammatical forms, as a verb, noun and adjective (*hygge, hyggen, hyggeligt*), she also confirms the centrality of light and home in *hygge*. A home *should* be cosy in order to be a home, and a killjoy who interrupts a *hyggelig* situation or refuses to let *hygge* take place is called a *lyseslukker* (light extinguisher).

When I interview people about their use and perceptions of various lighting technologies, they most often talk about *hygge*, homeliness, energy consumption and the 'cold light' from energy-saving light bulbs, antithetical to a homely, and by implication cosy, atmosphere, compared with the 'warm light' from incandescent bulbs. All informants distinguish between 'practical light' and 'cosy light', reflecting the activities conducted, rather than the specific room they are conducted in. Light is used to make the world visible, but it is also used to shape a desired atmosphere through which activities take place. In this way, informants change the 'feel' of their home during the day as the natural light changes or as their own mood (*humør*) changes, by turning on and off the electrical lights, drawing the curtains or lighting candles. In this luminous atmospheric orchestration of spaces, *hygge* is a central term, but it is important to note that it is by no means the only atmosphere opted for, even if it is the only term that has such a widespread application. In fact, most often people perform activities at home without any mention of *hygge* at all.

Hygge mostly implies dimmed, 'warm' lighting (c. 2700 kelvin) dispersed across the room to carve out smaller spaces within spaces around or in which to gather, often (but not always) assisted by candlelight – called *levende lys*, 'living light' as opposed to artificial. It is the visual contrasts that shape the cosiness that many Danes call for. Malene Lytken (2016: 93) has traced the roots of the term *hyggebelysning* ('cosy lighting') to 1948, although as she notes the visual norm it describes extends much further back. It was coined in a 'House wife Encyclopaedia' to describe the role of light in 'contributing to a pleasant or festive atmosphere in the room when resting, conversations, socializing or the like' (Lytken 2016: 95). The term *hyggelys* is widely used to establish the inseparable links between lighting and *hygge* as atmosphere, despite the definition of what it covers being quite wide, basically indicating soft, dimmed lighting.

These variations in luminosity guide people into specific places in the room, for example not switching on the light above the dinner table when guests arrive, in order to guide people to the sofa. Light, particularly the flickering flame, becomes a way of visually separating and carving out spaces that are not physically separated by, say, walls. Even in the daytime, candlelight is constantly used at lunch tables or at home to create personal, cosy atmospheres and offer focal points with nuances of glow, contrast, smell, air quality and warmth; that is, as a way of personalizing a space and indicating the ideals of *hygge*. Cosy light, in essence, is integral to the identity of the domestic realm as a *home* – an inseparable relationship between the particular quality of light and identity of a place that Tim Edensor and I have dubbed *lumitopia* (2017).

It is common to distinguish between light and darkness. Yet it seems quite clear from the many nuances in lighting, that firstly, there is no binary opposition between light and darkness, secondly, that while there is difference between dark and darkness, there is not common word to make that distinction in terms of light. Böhme (2017: Ch. 20) similarly critiques this opposition, and instead promotes the notion of *lightness* to emphasize qualities of light, to mimic the distinction between dark and darkness. As he notes, 'We see lightness not as an object but as a quality of the space that surrounds us' (2017: 199). While the notion of *lightscape* characterizes the scientific object of quantity and distribution of light in terms such as *kelvin, ra* and *lumen* and its distribution, *lightness* on the other hand promotes space as experience. This entails that the important point from a human perspective is not space as surfaces illuminated by beams of light through the spread of a lightscape, but as the embracing luminous quality of a space that surrounds us with glow, glare, brightness, etc., thus withdrawing the 'objectivity of light' to promote luminosity as an atmospheric element.

Tea lights, candles and flames from a wood-burning stove offer a diffuse and elusive materiality that shapes the room in different ways from electrical light, tying it in with ideas of intimacy and confidentiality. Several authors and scholars have commented on this: As Nørretranders and Eliasson note, 'you remember with your body how *hyggeligt* it is when you dim or turn off the evening light and light two candles on the table [. . .] Candles are also so beautiful because they create a bobble of light in the middle of the half-darkness [*halvmørket*]' (2015: 82, my translation). Although not in a Danish context, Gaston Bachelard's poetic note supports this: 'Reveries of this faint light will lead us back to the wee space of familiarity. It seems that there are dark corners in us that tolerate only a flickering light' (1988: 4; see also Bollnow 1963: 226). It may be an exaggeration, yet Edwin Ardener even claims that, 'in order to activate Danish domesticity, there needs to be candlelight' (1992: 28, my translation).

Variations exist as to what constitutes the particular character of cosy lighting, but the general setting is a dispersed, subdued light, mostly including candlelight, whereby light in essence becomes closely linked to understandings and practices of how to be and feel at home. As Hansen also notes:

> In its aspects of cosiness, of relaxed enjoyment in the warm aura of friendship or in the familiar security of kinship, the value placed on hygge reflects this desire for closeness, physical and emotional. The chandelier which lights the area in which the clustered group sits is commonly the sole or the strongest light in the room, visually circumscribing the cluster and defining it *as a unit.*
>
> HANSEN 1976: 59 my italics

There is, then, a particular care and appreciation for dimmed visual comfort. Some informants, of course, challenge *hygge* for being too stereotypically Danish or enforcing an atmosphere not warranted for. Yet except for my informants with immigrant background explored in Chapter 6, no one would light up a room with bare light bulbs, but always orchestrate it to some degree towards *hyggelys* by shielding off, or subduing, the direct glare.

The contemporaneity of meaning and presence

Susie's use of the tea light was an act of defining the unit; however, she goes further when proclaiming that the tea light *is hygge.* To follow Levinsen, she is making 'the common mistake of explaining *hygge* with reference to "candlelight" [which] reflects the lack of semantic-conceptual awareness' (2012: 108). According to Levinsen, the symbols of *hygge* closely associated with 'warmth' (the blanket, the flame, the hot drinks, etc.),

> do not define *hygge*, but they do tell a story of *hygge*'s ingrainedness in Danish culture. The message of 'candlelight' is Danish-specific because it relies on the Danish concept of *hygge.* The symbolic salience of 'candlelight' is to be found in its *hygge*-enhancing and *hygge*-augmenting functions.
>
> 2012: 109

Levinsen's exploration of *hygge* is about understanding meanings and the relationship between signifier and signified from an etic perspective. But perhaps this could be challenged if one approaches it with the 'purposeful naïveté [. . .] to take "things" encountered in the field as they present themselves' (Henare et al. 2007b: 2) as mentioned earlier. What Susie highlights is not so much relations of meaning of *what* the light is, but rather *how* it is, by shaping a presence of the world. She is blurring any clear distinction between meanings (representation) and the presence of things (presentation) where concepts present themselves in things without requiring interpretation as they are transformed into meaning (cf. Gumbrecht 2004: 80–1).

To further elaborate, another informant offering clear statements is Catherine, who pointed to this plural ontological nature when commenting that 'Candles are *hygge.* They create *hygge*', whereby cosiness is articulated as an *emergence* erupting in the simultaneity of cause and effect, it is *hyggelys* (literally 'cosy light'). It is used to denote the kind of light used in a cosy atmosphere, to distinguish light from the many other material and behavioural aspects that make up *hygge. Hyggelys* ontologically conflates the concept of *hygge* with the material phenomenon of light whereby a lit candle – or a subdued, dispersed light – is an invitation to, as well as symbol and instantiation of, *hygge.*

To understand this I look towards Edwin Ardener (1992), who has called this plurality of causality, essence and meaning expressed by Susie and

Catherine, a 'contemporaneity' (my translation of the Danish word *samtidighed*). It offers a perspective on human–object relationships where the parts do not lead to, or succeed, each other, but are present all at once: light is simultaneously a marker, an activator of action, a sign and a scene setter, offering a particular presence of the world (Ardener 1992: 27). In other words, the dimmed light simultaneously announces that the activity *hygge* should occur, while instantiating feeling in bodies, and creates an affective presence of things appearing in the luminous regime of a *lumitopia*.

Time and cosiness

Two things that have great impact on the atmosphere of a place are the season and the weather. Winter, with its short days in Denmark, shapes completely different atmospheres from the long summer nights. One informant noted that, 'if it is winter and cold outside, we need the candlelight to really make it *hyggelig hyggeligt*'. Yet, the term is also used in conjunction with other terms to describe different kinds of related atmosphere. Another informant, Margaret, for instance, described an introvert or 'intimate *hygge*', with drawn curtains, shielding off the apartment from the outside world, and light candles, creating something akin to a warm cave. This was opposed to the extrovert 'social *hygge*', which she found exemplified the first days of spring, or summertime, when she could hear people in the local park, neighbours talking on their balconies, and hear other people through her doors, opened to let the sunshine in. She also described a 'Christmas *hygge*' with all the garlands of light and the darkness that embraces the subdued interior lighting. As a consequence, different atmospheres depend on different kinds of light even if they all aim to make it *hyggeligt* in some shape. One kind of light is shed when having a cosy dinner with guests at the dinner table, another when having coffee on the sofa, or watching television. Yet all such different atmospheric experiences and material orchestrations go by the same term: *hygge*. In essence, we are thus dealing with a central term covering a wide spectrum of discernible atmospheres, yet defined in vague terms.

For my informants, *hygge* can erupt spontaneously in a moment of relaxation, informality or intimacy, but it does not necessarily equate with relaxation: You may relax without *hygge*. It does not have to be in the company of others, but could just as well be felt when alone. One informant stated that 'maybe you have devoted two hours to be together [. . .] but you forget time in those two hours. So, I think the most important factor is to forget time.' In this sense, *hygge* is both a spatial and a temporal categorization: 'It can easily be both a light or a room that is *hyggeligt*, as well as a situation with people,' as the informant Elina stated.

Hansen notes this centrality of time when stating that 'it is commitment par excellence to the present moment [. . .] "Hygge rushes in of itself as soon as one is carefree"' (Hansen, 1976: 54). Levinsen likewise notes how *hygge* has been described as the 'social nirvana in Denmark' where 'in order to have *hygge*, one should forget about time and place and focus on *nuet* "the now; the present moment"' (Levisen 2012: 92). As well as absent time, my informant, Marie, a geography student, contemplated that the perishability of candlelight actually instils a sense of time as the candle shortens as the evening progresses, unlike electrical light. In that sense, 'time is something practices "make" [. . .] Experience of time are part and parcel of the experience of practice' as Elizabeth Shove et al. found (2012: 129).

While atmospheres are commonly understood as a spatial phenomenon, this section has shown how not only spatiality, but likewise temporality and anticipation, are central to the atmosphere classified as *hygge*.

The power of vague definitions

By now it should be clear that *hygge* is about creating and experiencing a relaxed atmosphere where light and time are important aspects. Susie may have merged light and *hygge* when she approached the table, and her statement is seductively clear for the anthropologist to notice and introduce an academic argument. But once in a while I also meet less informative informants, such as Anne, who despite taking great care to decorate her home with personal items, also was very vague when it came to definitions. I asked her about what is so *hyggeligt* about candlelight. Her answer was less informative,

> I don't know. It . . . I guess I can't really explain. It's just something *hyggeligt*. I don't know. I can't put a word on it, other than it is *hyggeligt*. It gives such an . . . I don't know . . . intimate atmosphere. Or something like that. I don't know.

There is a difference between not knowing and being vague or ambiguous. Anne could not verbalize an answer, but she would know in a very bodily sense. For most other informants, it was not about *not* knowing but about having several definitions or lacking the ability to verbalize the knowledge. Susie's statement, about the arrival of light as *hygge*, is not just an expression about how the world *is* in that exact moment, but also an expression of *how* the world is present and what it *should* be – it presents an interest of perception, entangled in social evaluation of the nature of a social gathering in Denmark. Despite such clear statements about light as *hygge*, further questioning of my informants about the nature of *hygge* causes a shift between clarity and vagueness to appear, for instance when Nanna, a mother in her mid-thirties, states: '*Hygge* is many things, but of course, we say that candlelight is also *hygge*. I guess it is. Maybe it is also that. They set that kind of atmosphere. It is something, yeah, something you gather around.' She goes on to say:

> It is like I mentally tell myself that now I want to make it *hyggeligt*, so I turn on this light. I guess it is like that. When I do this, it becomes *hyggeligt*. But it doesn't. It may not lead to *hygge* – I also know that. But the frame is sort of set for it. The stage is set by lighting the candle.

On the one hand, Nanna points to the way that candlelight instantiates a sense of *hygge*. However, after further reflection, doubt comes to mind – 'I also know that it is not like that' – after which light becomes a requisite for setting the cosy atmosphere. As Levinsen notes, there is an underlying assumption,

> that candlelight, as a matter of fact, *are* transmitters of *hygge* [. . . but . . .] 'candlelight' does not define *hygge*. In fact, it works the other way around: the true Danish message of 'candlelight' can only be understood via the concept of *hygge*.
>
> 2012: 107, 108

The informants navigate a cross-field of clarification and doubt, interchangeably between a 'pathic moment' – of a pre-conceptual presence of being seized by the atmosphere where light is considered *hygge* – and a 'gnostic moment' of conceptualization and classification of the perceived, where light may lead to *hygge* (Strauss 1966: 10). As shown in the discussion of causality and definitions of *hygge*, there is a pressing need for 'mediating and filling the yawning gap between comprehension and affective involvement with a smooth terminology by conceptually embracing unmediated experiences of life' (Hermann Schmitz quoted in Riedel 2015: 86). Informants *sense* that light instantiates *hygge*, but at the same time, when further reflecting upon it, they *know* that it is not. Therefore, sensing is not only a mode of knowledge but also an affective relation through an atmosphere (see also Barbaras 2004; Strauss 1966).

Even if vaguely defined, Levinsen concludes that *hygge* 'is a mental ontology, a concept which is "real" to people' (2012: 91). *Hygge* then is an emic term for a particular atmosphere: it is real, it is what is experienced, yet it also escapes clear definition. However, if one follows Levinsen's approach of bypassing this realness by separating the symbolic and semantic properties from the materiality of the candlelight or just dimmed light, there is a danger of overruling emic understandings of the atmospheres that life is lived through. As Holbraad warns: 'Instead of treating all the things that your informants say of and do to or with things as modes of representing the things in question, treat them as modes of defining them' (Holbraad, 2011: 12, cf. Bijker, 1995: 77). *Hygge* is light, if one takes the informants' words at face value.

But informants also state that this is not the *only* case, showing that there exists a multiplicity – both *hygge* and light are also something else. 'Light is *hygge*, or maybe *hygge* is not the proper word', as several of my informants stated. *Hygge* is, after all, a comfortable word to use, sometimes a metaphor, but often used in need of a better term for the plethora of situations it occurs in. *Hygge* denotes multiple things and phenomena (even in the same grammatical form): tangible things, modes of appearance, ways of being together. In other words, rather than an inseparable *part* of a thing, *hygge* becomes an affective condition – an atmosphere – inseparable from and achieved *through* light, ontologically conflated as *hyggelys*. To the above informants, light and *hygge* are both the same *and* different things – conflated and related – and neither the unification nor the difference is something the informants are completely decided upon, leaving the researcher in a world where vagueness is a premise of the world rather than an outcome of a flawed methodology.

Conclusion

Hygge, in essence, is a way of being together in Denmark; a goal in itself for *how* practices and perceptions should unfold. It may include light and many other material objects that spread cosiness as a (temporary) affective property of a thing, but essentially *hygge* is about relaxed and informal atmospheres, absent apparent hierarchies. The relationship between light and *hygge* is not one of either/or, but of simultaneously building an emerging, rather than causal, atmosphere. The undecidedness as to the nature of the relationship between light and *hygge* is anchored in this double position of being both object *and* not object.

This chapter took Susie's proclamation that light is *hygge* as a definite statement about reality. One perspective would be to claim that it is the way Susie performs and talks about light that defines what it is, and her statement was just one out of many different versions. On the other hand, I wanted to push the analysis and talk about the way in which the material environment becomes embraced by an atmospheric tone. It is important to note that *hygge* is by no means the only atmosphere(s) people opt for, or even one they explicitly mention if conducting activities that may encompass elements of *hygge*. It is nonetheless an important aspect of Danish domesticity. After this clarification of how *hygge* unfolds in oscillations between clarity and vagueness, as well as both object and concept, we can see that, in essence, it describes the feeling of a space, emerging between the (temporary) affective properties of material phenomena and the goals of social interaction. On this ground, the following chapter turns to a more conceptual exploration of what constitutes an atmosphere.

4

Atmospheric realities

Introduction

Think of the marvel of entering a museum or cathedral; an uplifting early spring afternoon in the sun; a hectic marketplace; the solemn silence of a snow-filled forest; a romantic dinner; being stuck in traffic; cheerfully singing along at a concert; the confusion emerging after a terror attack; a depressing urban space. It is an immediate sensuous experience of both stepping into a place and being part of it, not necessarily good or bad, but still experientially there. You may later *think* of it without being able to put into words what it was that made it that way, even if you recall the bodily sensation. Lack of term, does not, however, make it any less real. One such example is 'the eerie, forlorn atmosphere of a place that's usually bustling with people but is now abandoned and quiet',[1] recognizable to many, but which only recently has been defined by the word *kenopsia*. In the growing literature on atmosphere, this non-discursive nature is commonly illustrated by such 'think of' exemplifications to show how quickly one gets a feeling about a space through the very affective sensibility of the body as tacit knowledge: the feeling of how hate/love/fear was 'in the air' (Brennan 2004: 1; Hasse 2012; Johansen 2015; Zumthor 2006: 13). One is both *part* of an atmosphere as a co-producer, and may enter *into* it and become captured by it. Stepping into a funeral may dampen a joyful or exhilarated mood, and anxiety may disappear when you meet people at a party. What you feel is not just your own attunement but also a tinge of other people, and the affective presence of the material world that embrace and alter the experience of the world. We are part of the space and the space is part of us.

The *becoming attuned* nature of engaging atmospheres is important to stress here. The atmosphere is not just inside *my* head when I enter spaces, but also something the surrounding people experience, take part in and transform. People come with anticipations, moods and social values, but atmospheres are not just a product of the human psyche. Architecture, for instance, is a material form that deliberately seeks to direct the atmosphere

(Bille & Sørensen 2016b; Borch 2014; Pallasmaa 2014a; Wiil 2016; Zumthor 2006). William Lam noticed almost 40 years ago that, 'Light has always been recognized as one of the most powerful formgivers available to designers, and great architects have always understood its importance as the principal medium which puts man in touch with his environment' (1977:10). However, atmospheres cannot be reduced to the material qualities of the infrastructure either, as also shown in the previous chapter on *hygge*. Atmospheres change during the day according to sensorial inputs, the people and the activities taking place, as illustrated with the summer-*hygge*, intimate-*hygge*, etc. in Chapter 3, even if this transformation of atmospheres has lacked detailed focus in empirical studies (see Edensor 2015b).

Atmospheres and other related terms

One of the key figures in exploring atmospheres is Gernot Böhme, as already illustrated in the introduction. In an oft-cited passage, he describes atmospheres in the following way:

> Atmospheres fill spaces; they emanate from things, constellations of things, and persons [. . .] Yet they cannot be defined independently from the persons emotionally affected by them; they are subjective facts. Atmospheres can be produced consciously through objective arrangements, light and music [. . .] But what they are, their character, must always be felt: by exposing oneself to them, one experiences the impression that they make. Atmospheres are in fact characteristic manifestations of the co-presence of subject and object.
>
> 1998: 112–14

With Böhme taking the lead, there has been a recent proliferation of conceptual discussions of atmosphere (Anderson 2009; Bille et al. 2015; Böhme 1995, 2001a; Gandy 2017a; Griffero 2014; Rauh 2012), which allows us to better understand the specific affective qualities of architecture (Pallasmaa 2014b; Zumthor 2006), the home (Bachelard 1994; Bille 2015a, 2015b; Daniels 2015; Hauge 2015; Olesen 2010; Pennartz 1986), the event (Anderson & Adey 2011; Borch 2012; Edensor 2012) and urban, commercialized or public spaces (Duff 2016; Ebbensgaard 2015; Hasse 2012; 2008; Healy 2012; Schüll 2012; Thibaud 2011).

Böhme was by no means the first to discuss atmospheres, however. He very much engages with the pioneering work of Hermann Schmitz (e.g. 2014; 2011), and Michael Hauskeller (1995) came out with a similar argument in the same year as Böhme's (1995) widely recognized book *Atmosphäre*. But Böhme has been particularly dominant for his clear argumentation, which has been translated into English (1993, 2017). Even earlier works include Hubertus Tellenbach (1968) seminal work on taste and atmosphere, which delves into the particular affective sense of being in space in an attuned way, while Christian Norbert-Schulz discussed the affective power of architecture as *Genius Loci* (1980). Furthermore, discussions of atmospheres also at times seem to mimic the discussion of *Stimmung* presented by Martin Heidegger (1996: 126), and later on by Leo Spitzer (1942a, 1942b) or Otto Bollnow (1941). Bollnow describes spaces as 'tinctured' by the presence of people and things (1995: 33), yet note here how he describes *Stimmung*:

> it is also wrong to solely put the *Stimmung* on the subjective side and to assume that it then, in a way, rubs off on the world [. . .] *Stimmung* is therefore also not assigned to the isolated 'inner life' of human beings, yet human beings are included into the whole of the landscape, which in turn is not something that exists separately, but is rather in a particular way referred back to human beings.
>
> 1941: 39–40, my translation

It appears that clear distinctions among concepts are often somewhat obscure or inconsistent. For instance, *Stimmung* is often translated as attunement or mood.[2] It is seen by most in the German tradition, drawing on Heidegger, as the 'basic affectivities' of human existence (Bollnow 1941: 34, translation of *Grundbefindlichkeiten*), compared with the spatial focus of atmospheres. However, Böhme, concurrent to this 'basic affectivity' perspective, sees *Stimmung* as a special category of an atmosphere, distinct from the communicative, social, synesthetic and motion-impression categories he also defines (2001b: 89–90).

Maybe there are inconsistencies across different authors, but one of the problems is that most often the discussion of the related terms atmospheres, *Stimmung* and ambiance, or even Peter Sloterdijk's work on spheres (e.g. 2016), is that they are solidly based in a French or German context, which still is only sporadically translated. This makes Matthew Gandy warn that, 'one of the difficulties with the way the term "atmosphere" has been adopted, especially within Anglo-American cultural geography, is the lack of robust historiographies' (2017a: 368). Luckily, within the last few years, more and more of this French and German work is being translated, summarized, and applied in cases making such similarities appear and discussed (cf. the journal *Ambiances*, and Bille et al. 2015; Borch 2014; Gandy 2017a; Griffero 2014; Hasse 2014; Thibaud et al. 2012; Thibaud 2002, 2011, 2014, 2015; Riedel 2019).

I will not further discuss the difference and similarities in academic terminology, whether one prioritizes spatiality or individual person with *Stimmung* rather than collectivity with atmosphere, or puts emphasis on fixating affect in architecture I will rather relate through the term atmosphere to the central issue of understanding the sensation of being in space, and its socio-material premises and cultural logics.

Vagueness and the body

Atmospheres are an integral part of being human, and not necessarily something that can be verbalized or easily defined. Atmospheres are spatial phenomena, different from a table. However, this does not make them any less 'real'. They must rather be seen as quasi-objective half-things like the weather or the wind (Böhme 1998b: 114, 2001b: 59, 2006b: 16; Schmitz 2014: 39). They 'exist' as vague entities even if they have weak or strong intensity, and negative, positive or insignificant influence. The task for an atmospheric description is thus to designate 'a vague entity in a precise way' (Griffero 2014: 7), as I have elucidated above in terms of *hygge*, or any of the other atmospheres that people sense, such as uplifting, tense, serene, etc. The fact that atmospheres are unlocalisable may seem like a vague and ungraspable point of departure compared with the concreteness of studying material objects, speech, acts or actions. Yet, as Griffero notes, 'why on earth, in fact, should solid and contoured bodies be more real than vague entities, which we experience without referring them to solidity, such as fluids, gas processes or even quasi-things like atmospheres?' (2014: 10).

Central to an atmosphere, on both an experiential and analytical level is the role of the body. Böhme constructs the notion of *the space of bodily presence* as 'the manner in which I myself am here and am aware of what is other than me' (2003: 5), which may be inspirational to our understanding here, as it erupts in what he calls the spaces of actions, moods and perceptions. He argues that,

> The space of moods is the space which, in a sense, attunes my mood, but at the same time it is the extendedness of my mood itself. The space of actions is the space in which I can act, but also the scope of my possibilities. The space of perceptions is the space in which I perceive something, but also the expansion of my involvement with things.
>
> 2003: 5

The focus on the body, and its means and motivations to move, perceive and be attuned, is pivotal in an anthropological understanding of atmosphere as unfolding through, and impacting on, people's lives (Stewart 2011). Yet there is also a more normative and judgemental aspect in terms of understanding the particular kinds and characters of atmosphere that people opt for, for instance *hygge*. Atmospheres are *what* is felt in a space but also what *should* be felt and thus entangled in practices of making a place 'feel right'. That is, there is a sensation of atmosphere but also a process of shaping its character.

While Böhme gives little detailed attention to the specificity of what he calls the social conventions (2001b: 89–90, 2006b: 124) such as the cultural, historical and social aspects of atmospheres, I follow Böhme's conceptualization in so far as understanding atmospheres as the 'co-presence of subject and object'. This allows us to focus on both the organization of material phenomena and the cultural values they unfold through. Atmospheres emerge in the process of shaping the co-presence of the affective properties of material culture and attuned people as 'tempered spaces' (Bollnow 1963: 230). Furthermore, while it has been claimed that the phenomenological approach to atmospheres is ill-equipped to explore the material and political aspects of for instance lighting (Gandy 2017b: 1101), Böhme actually continuously insists that attending to atmospheres also offers the potential of critique of contemporary staging of aesthetics to reach political or economic goals (2017). In essence, according to Böhme, people should learn or be taught to be more aware of atmospheres, as they may otherwise creep in and manipulate people without them realizing it.

With the ability to both stage, reflect on, and be manipulated by atmospheres, it also becomes clear that one may distinguish between the 'experienced atmosphere', what a place feels like, and the 'reflective atmospheres' as the shared conceptions about what the specific quality of an atmosphere entails as it is addressed by thought (Thibaud 2014: 282). It is thus important not to ignore cultural classification and appreciation of atmospheric experiences. For instance, one may experience a relaxed atmosphere that is talked about and conceptualized in terms of *hygge*. Hence, what may be perceived as *hygge* by a Dane may be thought of as a romantic or sleep-inducing atmosphere for people with another cultural horizon. Atmospheres thus have culturally informed affective potentials grounded in the meanings, practices and presences of people and things.

From atmospheric sensations to practices of attunement

Grounding discussions of atmosphere in people's lives may offer more nuances than those mentioned above. Turning to our Danish case, two emic words are used interchangeably to describe what I already denoted as 'atmosphere':

stemning and *atmosfære*. Mostly informants talked about *stemning* such as a *hyggestemning* (cosy atmosphere) or *feststemning* (party atmosphere), whereas *atmosfære* is less often used, although still frequently. Anna was one of my more articulate informants when it comes to describing her sensation of space, and her perspective resonates with many other informants. To Anna, *stemning*,

> entails that there are also other people. That there are some people to set the *stemning* (*sætte stemningen*). An *atmosfære* on the other hand can be shaped by the things present. So, there does not necessarily have to be any people, like, doing it. An *atmosfære* could be like walking into an old abandoned warehouse. You can find a really special atmosphere there.

In other words, if one were to draw more general distinctions based on such statement, the co-presence of objects and one subject is seen as atmosphere, while the co-presence of objects and more subjects is seen as '*stemning*'. The *stemning* emerging from the Danish informants is a subjective derivation of an atmosphere (cf. Böhme 2001b: 47) that 'seems to exist everywhere when "people are with other people"' (Levisen 2012: 96–9). It is the awareness that some sort of atmosphere, good or bad, 'rules' the place as a social fact. There may simultaneously be another understanding of *stemning* with inherently positive connotations as a social potential, 'when people are willing to give themselves over to a socially created feeling' (2012: 99). *Hygge* is described as such a *stemning* by my informants, yet as Levinsen clarifies,

> it is more appropriate to view *hygge* and *stemning* as words which supplement each other in describing aspects of Danish togetherness [. . .] Together, *hygge* and the concepts of *stemning* make up key Danish points of orientation in social life. They all emphasize [. . .] the social emotions emerging from the dynamics at play where 'people are with other people'.
>
> 2012: 96, 99

In essence, the two notions of atmosphere in Danish are defined by the presence of other people (*stemning*) or not (*atmosfære*). Yet there is more to it, if we try to understand the feeling of a space in terms of attunement and atmosphere.

There are two different emic kinds of *stemning* in Danish, beyond the distinction that Levinsen makes between *stemning* as a social fact and as a positive social potential. In one version, it is a *spatial term*, equivalent to Levinsen's definition. The other deals with the processual and existential nature of *stemning* as an *attunement*. *Stemning* is etymologically linked to the German *Stimmung* and I here draw inspiration from Heidegger's notions of *Stimmung*, discussed above, in the sense that it is an emergent mode of existence through which the world is disclosed to people – 'how one is and is coming along' (1996: 127), rather than a psychological state of mind. It comes neither from the inside, nor from the outside, but arises as a way of being in the world (cf. Elpidorou & Freeman 2015; Ferreira 2002; Harman 2007: 68). In that sense it is also different from a feeling, in that it is not necessarily directed towards something, being 'angry *at*', for example. For instance, 'I am not in *stemning*' (i.e. attuned, *jeg er ikke i stemning*) indicates that people are either attuned in a particular way (*være i stemning*) or can become attuned (*komme i stemning*) – or not – to do certain things, and you may deliberately design a space to such attunement (*sætte en stemning*, set an attunement).

Attunement is thus a *process* of inhabiting the world compared with atmosphere as a *temporal* condition of spatial encounters (see Pérez-Gómez 2016 for a discussion in relation to architecture).[3] To understand this, one needs to see how an atmosphere is a 'spatial carrier of attunement' (Böhme 1995: 29, my translation), merging elements of other people's attunement and the affective presences of material phenomena (including weather and the intangibility of light). A space, in this sense, does not *express* an attunement but *possesses* it (Griffero 2014: 61) – although again it is important to note that there is a temporality to this possession. Even if the atmosphere may capture you, your attunement may also tinge the atmosphere. Informants noted that a space may be orchestrated to *hygge*, and that one may also feel relaxed and in a *hyggeligt* mood. But one's thoughts may also wander, of course, interfering and shaping one's attunement. In this line of thinking Kathleen Stewart talks about atmospheric attunement as 'an intimate, compositional process of dwelling in spaces that bears, gestures, gestates, worlds' (2011: 445).

When informants state that one type of atmosphere, *stemning*, relies on the (always attuned) people present, what is sensed is a derivation of their attunements. The other version of atmosphere, *atmosfære*, is the way an attuned person is embraced and may become attuned by the affective presences of material phenomena. Atmospheres, in this line of thought, are 'out there' in the spaces '"tinctured" through the presence of things, of persons or environmental constellations' (Böhme 1993: 121), where attunement is a way of being in the world, more than a psychological mood or feeling directed at a particular object or situation. While you may personally be attuned to a party, you may also have a spatial sensation of how a room feels depressing. The atmospheric imbalance is a potential for change, of the attunement of either room or person.

While the above conceptual discussion of atmospheres focuses on perception predominantly located in philosophy, much of the literature on atmospheres seems, from an anthropological point of view, to underestimate how people not only perceive the world through atmospheres, but also perform activities and practices and transform it (Bille et al. 2015; Jürgen Hasse 2008;

Hauge 2013b; 2015; Jensen et al. 2015; Thibaud et al. 2012). They light the home to start vacuuming, or as part of preparations for guests to arrive. In some phenomenological literature on perception – Merleau-Ponty's (1968) notion of lived experience in particular – it is quite clear that perception entails practice, even if the 'phenomenality of experience transcends the mere practice' (Ebbensgaard 2015: 116; cf. Simonsen 2007). As Jean-Paul Thibaud notes, 'perception cannot be reduced merely to passive contemplation of the world – it involves moving in a certain way' (2011: 210). In this sense, 'perception is active. It is formed as bodies move in space, and materiality and perception are intertwined in the process' (Cathrine Hasse 2008: 44).

A recurring argument here is then that people actively seek to attune themselves to the activities or the spaces they are in, through light, and to what they (want to) feel or do. Getting home, taking one's jacket off to feel at home, lighting a candle to remind oneself that one is no longer at work, taking a hot-water bottle into bed to make it cosy and warm. 'Light is not only practiced – it affects us' as Ebbensgaard argues (2015: 116). Bettina Hauge (2013b; Wågø et al. 2017) also argues that attunement is often about shifting roles, where practices are performed to create new atmosphere in the home, such as opening windows for air ventilation, light candles, which in a sense act as domestic rite of passage between work and leisure, the everyday and celebrations. Such an attuned appropriation of space is part of our ordinary practices and habits, even if people, of course, may not necessarily succeed in finding a balance between wanted atmosphere and ability to orchestrate the space and oneself to it.

People, spaces and things are continuously attuning one's bodily presence. One may seek to change the atmosphere, for instance by light, and the atmosphere, made up of the elements of people's attunement and affective properties of things may capture them by their embrace. It is a vagueness of the elements that allows for such oscillations between being seized and transforming atmospheres that make them potent analytical objects to understand our world through. In this sense, atmospheres are spatial carriers of attunement, either through the practices of attuned people or the affective properties of things.

Affecting presences

In the above I have alluded to atmospheres as tempered spaces emerging as the continuous co-presence, and co-productive potentials of subject(s) and objects. To sum up, I understand the former – the subjects – in terms of being in a constant process of attunement that guides practices and ways of seeing the world. To understand the latter – the objects – an understanding of the way the material world may affectively 'move' a sentient being is needed. This is also where the extensive literature from the material culture studies, also dubbed the 'material turn', becomes central. The focus on material culture essentially highlights the role of things in normalizing and moralizing human interaction. Material culture frames human life, allowing people to live through things and not just among them, and this is no less so with intangible material culture, such as light or sound (Hicks & Beaudry 2010; Miller 2005; Tilley et al. 2006).

In Robert Armstrong's seminal work on Yoruba art, he critiques a projectionist approach to things where people invest objects with meaning or symbolic content 'behind' the artefacts, since they fail to notice the artefacts' selfhood and what he calls their 'affecting presence'. Affecting presence denotes the ways things impose themselves on people, as a 'perpetuating, affecting act – a near-being, with its unique "personality" continuously asserting its own existence' (1971: 29). To Armstrong,

> [t]he affecting presence is a thing-in-itself – a *presence* as I have called it here – and not a symbol because the creator does not build into his work cues to some real or imagined affective estate external to the work itself, but rather strives to achieve in that work the embodiment of those physical conditions which generate or are causative or constitutive of that emotion, feeling, or value with which he is concerned [. . .] The meaning of the affective presence – or more properly its 'feeling' – is thus not external to itself [. . .] An affecting presence is thus a presentational presence, and it is existentially 'all there'.
>
> 1971: 31–2

I have argued elsewhere (Bille 2017; Bille et al. 2015) that in an atmospheric perspective, a thing is articulated not as being-there or not-being-there, but instead as 'the ways in which it goes forth from itself' (Böhme 1993: 121). These ways of presencing or 'going forth' from itself are what Böhme terms 'the ecstasies of the thing', i.e. the way in which a thing qualitatively and sensuously stands out from itself (1995: 32–4).

For some readers this may resonate with a recent spate of literature on 'affect' (Anderson 2014; Massumi 2002; McCormack 2008; Navaro-Yashin 2012; Pile 2010; Stewart 2007; Thrift 2007, 2009), which conceptualizes affect as a preconscious 'flow' that attunes the body. But rather than an affecting presence of the objects in Armstrong's case, it is actually more reminiscent of the conceptualization of atmosphere as we have explored it above. The key idea in this literature is that there is a pre-conscious body where affects are felt and spread before any cognitive or cultural regimes are enforced – as when Brian Massumi claims that there is a half a second delay between sensation and cognition (2002). This literature on affect, rather than the affecting presence, has more recently been coupled with atmospheres as 'collective affects' or the pleonasm 'affective atmospheres' (Anderson 2009; Ash 2013; Buser 2013; Edensor 2010; Gandy 2017a; McCormack 2008; Stephens 2015) – suggesting that there are non-affective atmospheres, or that solitude makes atmospheres impossible. While immensely productive to understand the non-representational aspects of human lives, it has also been extensively criticized for being essentially sequential in dividing sensation and cognition into two separate fields, where one leads to the other in a sequence (e.g. Leys 2011). The problem is that other data have also shown how the affective engagement with the world is constantly framed and influenced by unconscious cognitive interpretations (Wetherell 2012). Affect and emotion becomes separated as distinct field, which, however, is criticized by Edensor (2012: 1105), for ignoring how such distinctions are rather elusive and amorphous. Furthermore, the relationship may not find similar meaning or distinction in all languages. Focus on affect thus leaves us with the problem of distinction between emotion and affect, and body and cognition, but also questions about how that affect is transmitted from one person to another (discussed in Bille and Simonsen in press; see a nuanced discussion in Anderson 2014).

Another critique could also be raised in terms of terminological usurpation. Navaro-Yashin, for instance, defines affects as 'hazy and atmospheric' (2012: 168), drawing on Teressa Brennan's notion that 'any inquiry into how one feels the other's affects, or the "atmosphere", has to take account of the physiology as well as the social, psychological factors that generated the atmosphere in the first place' (Brennan 2004: 1). 'Affect' in this particular conceptualization, as I have argued elsewhere (Bille et al. 2015; Bille & Simonsen in press), is, in essence, covering what the literature on atmosphere highlighted, although without making the spatial spread a defining feature. So if affect is atmospheric, but without the explicit spatial spread, what then is the point in talking about affect, as a noun, rather than, say, following Armstrong and focusing on the affecting presence?

Thus, while I believe we get a fuller understanding of lived experience in spaces from an atmospheric approach, what we however do learn from emphasizing the affecting presences is to see the material world around us as active and potentially attuning people in places. I stress 'potentially', since there is no determinism in whether it manages to capture the sentient being or not. The atmosphere is not a stable 'spirit of the place' (Norberg-Schulz 1980). People become attuned, and this attunement is mediated through the affective material presences, light included, applied in orchestrating atmospheres, and the practices that people take part in. The point is that things are affective, more than merely projecting people's emotions. They are active, and one such material, although intangible, object, is light, and how it may be staged to achieve atmospheric outcomes: how light takes part in attuning the atmosphere as a co-presence of affective material phenomena and attuned people.

Staging atmospheric light

FIGURE 4.1 *Theatrical lighting in Nordvestparken, Copenhagen. 'Everything is illuminated with aesthetic intent' (Böhme 2017: 203). Photo by the author.*

Böhme notes how light is known in the appearances of things (2013b: 136), in that 'light is in itself something atmospheric that can fill an entire room, that wraps itself around things and embraces them' (2013b: 135, my translation). As art historians, theatre directors and cinematographers will easily tell you, light plays a crucial role (though not exclusively) in efforts to shape people's experience of a place; to appreciate and manipulate light for its 'decisive importance in experiencing architecture' (Rasmussen 1964: 187). The creative designer may not be absolutely sure that people will get goose bumps or shed a tear, but they may certainly set the scene to elicit such emotions. At other times, minor changes to the light may completely change the atmosphere and appearance of an object. The way in which the world presents itself is toned by light settings that most people barely recognize, even if – according to some recent research – they guide moral judgement, creative processes and the physical responses of the body (Chiou and Cheng 2013; Czeisler 2013; Steidle and Werth 2013).

If a precondition of atmospheres is that they are felt, and defines a co-presence of subject and object, it raises the question: can they also be staged? Some have heavily criticized the idea of producing atmospheres (Rauh 2014: 247–8), while Böhme is more convinced in his writings, as he sees this staging being actively promoted in society through politics, media and design. He states, for instance, that they can be actively produced (e.g. 1995: 97, 1998b: 114, 2013a), yet elsewhere he also says 'without the sentient subject they are nothing' (2013b: 103, my translation).

One way to overcome such contradiction is to envisage the fact *that* they are as being something that may be produced or staged, but *what* they are is something that has to be felt by the sentient subject – a subject informed by cultural upbringing and the specificity of the social situation. This distinction between 'that they are' and 'what they are' is something Böhme defines as the difference between the 'atmospheric' and the 'atmosphere', where the 'atmospheric phenomena are distinguished from the atmosphere by the substantial absence of subjective moments' (2001b: 60, my translation). Designers and home-owners alike may orchestrate spatial properties through material objects but their quality unfolds in practical experience (Bille et al. 2015; Böhme 2013a; Sloane 2014). Although Böhme does not take this point further himself, these nuances between the 'felt' and the 'staged' reveal the role of atmospheres in social life. They mark the difference between what kind of atmosphere *is* felt when people are together (or alone), and what *should* be felt, which of course rests on very cultural ideas such as hospitality, intimacy, morality, power, gender, etc.

There is increasing focus on such atmospheric, sensuous and affective attunements in, or *as*, architecture (Bille & Sørensen 2016a; Borch 2014; Griffero 2014; McQuire 2005; Pallasmaa 2014a; Rasmussen 1964), and particularly with the advent of new LED technologies to illuminate urban spaces, light has a key role. As architect Juhani Pallasmaa has observed:

> Nowadays I don't regard architecture as a building in itself; it is a means of revealing something else. For me, light is the most ecstatic architectural experience there is, and in many ways the best architecture is a preparation for the experience of light.
>
> IN PLUMMER 2012: 13

Tim Ingold takes this focus on light and atmosphere even further by stating that all living architecture is atmospheric. Understanding architecture without looking at atmosphere is like trying to understand music by only looking at the instruments (2016). There is no separation between architecture and atmosphere, when the focus is on how the world presents itself in lived experience. This focus on the lived experience – a central feature of phenomenology – seeks to overturn a distinction between nature and culture, object and subject, by exploring everyday perceptions (Csordas 1994; Desjarlais & Throop 2011; Jackson 1996; Tilley 2004; Zahavi 2003). In such a perspective, luminosity – or sound or smell for that matter – is not a 'secondary feature', but is integral to the object; a thing is serene because the atmosphere it radiates brings a person to a serene mood (Böhme 1995: 34).

To illustrate his approach, Ingold uses the weather as an example of the medium of perception: you do not observe the weather but you inhabit it as 'an experience of light itself' (2005: 97). You do not hear the sound *of* a wave crash upon the beach, but you hear the wave; you do not see the light *from* the sky, but the sky, etc. (2011: 134). What I take from Ingold is an awareness of the integral ways in which light (and other elements) shapes the qualities of the spaces inhabited, letting you live them 'from the inside' (Ingold 2000: 265, 2011: 134; Merleau-Ponty 1964: 178). One could then argue that what informants experience in the most ideal moments of *hygge*, is not light or a room that is *hyggelig* (cosy), but *hygge* (cosiness) itself. It is the experience of a *lumitopia* where light and place are integral to the atmospheric character experienced.

The nature of atmosphere cannot be limited or even pinpointed materially or geographically, and yet it has an ontological reality that tones the world and attunes people. Atmospheres may take the form of intensities that provoke an interest of perception about how the world should be, whereby an object, like Susie's tea light in chapter 3, is momentarily confined in a contemporaneity of matter and concept, only to separate into entities again as the atmospheric intensity drops. A space thus does not contain an atmosphere as a property of the space, but it holds a temporary possession of atmospheres as the entanglement of attunements and the affecting presences and ecstasies of things.

Ontology of elements

Such an atmospheric approach has epistemological ramifications. First and foremost a focus on atmospheres means a shift from what a thing *is*, or is enacted to be, to a focus on *how* the world is presented as it is. People do not experience the world objectively as a 'factual fact' (*Realität*), but atmospherically as 'actual fact' (*Wirklichkeit*) (Böhme 2001a: 57). Both *Realität* and *Wirklichkeit* are translated into 'reality' in English, yet the distinction highlights how there may be a difference between the calculations and reduction of the world into parts, and how the world is experienced. This shift forces a move from the ontology of objects, i.e. each individual thing, to the ontology of the elements where 'the element is not subjective, nor is it what is perceived. It is the dimension through which perception takes place' (Barbaras 1998: 222; cited in Thibaud 2015). As Böhme notes,

> The primary 'object' of perception is atmospheres. What is first and foremost perceived is neither sensations nor shapes or objects or their constellations, as Gestalt psychology thought, but atmospheres, against whose background the analytic regard distinguishes such things as objects, forms, colours, etc.
>
> 1993: 125

In essence, experience of the material world becomes what it is to the observer through the atmospheres that are created in the spatially extended *inbetweenness* of affectively tinged matter and attuned subject. Perceiving thus 'involves not only discerning objects in the environment, but *experiencing the state of the medium* at a given time' (Thibaud 2011: 212, original emphasis). This is an addition to the last 25 years of sensory studies (Howes 2004, 2006; Pink 2009; Seremetakis 1994; Stoller 1989), which have shown us how the priority, primacy and indeed the very categorization of senses are entangled in cultural worlds. The atmospheric approach, of which the notion of *hygge* is an exemplary case, aims to focus on how such sensations are spatially extended in affective encounters with atmospheres, as the co-presence of subject and objects. Important to note here is that such co-presence does not entail symmetry of equal impact; the character of the co-presences must be anchored empirically. The task in this chapter has been conceptual clarification. The aim of the following chapter is to anchor such discussion.

Conclusion

It may be difficult to describe the atmosphere precisely in any given situation, and yet, as social life takes form, atmospheres, with all their lack of clarity, also structure what the world *should* be. Susie's proclamation in the previous chapter that 'here comes the *hygge*' when bringing a tea light, was not so much a statement about the ontology of the tea light, or the emic concept of hygge as cosiness, as it was about the ontology of the elements – the atmosphere that *is* and *should* be emerging with the light. With a focus on the ontology of things, an object is either one or the other, enacted by the person at a certain time. In the atmospheric focus on the ontology of the elements, it is, by contrast, the contemporaneity of being both clear and vague, of being simultaneously meaning, presence, sensation and action that is the point. An atmosphere embraces its surroundings by permeating them with a tone. As subjective facts, atmospheres are the contact zones of *inbetweenness* that cannot be reduced to the object or subject, but are always there: not only *in* the relation but also *as* the relation between people, places and material phenomena. What is left is the atmospheric reality of such vagueness, which emphasizes the importance of co-production, and the process of attunement. In other words, the focus is on *how* the world came to appear as it does, rather than what the world *is*.

5

Atmospheric communities

Introduction

In wintertime, the sun sets at around 3.30 p.m. on Islands Brygge, a residential area of the harbour close to central Copenhagen. As darkness descends, I can see from my apartment how the neighbourhood is slowly illuminated and comes to life. First, the staircase light, which automatically turns on as people enter it, then as they enter their home, the light in the hallway is lit, then the kitchen, the living room and the other rooms. Christian and Hanna, a couple living on one of the lower floors, turn on all their lights when they come home: 'It is a way of coming home,' they explain, as an affective, more than spatial journey, reminding them that they are no longer at work. They feel that the apartment is very small, with only little natural light coming in, so turning on all the lights is a way of enhancing the space and making it feel homely. Elsewhere, as the evening progresses, candles are lit, a variety of lamps are turned on and off in different rooms, staying lit even when there are no people present. Sometimes a light is on in one room simply to spread light into an adjacent area, where the installed lighting would be too bright – for instance turning the lights on in the kitchen in order to illuminate the hallway, which would otherwise disturb the darkness in the children's room. Lighting is about caring. Most apartments are similarly visually augmented by the shifting coloured light from the television Later, the light is turned off in each room as the evening draws to an end and the light in the bedroom dims. As night falls upon the neighbourhood, only the moon, reflecting light from the clouds, street lights, cars and sporadic light from light sensors in the staircases of those returning home late, illuminate the neighbourhood.

Much the same lighting practices occur in summertime, although the role of artificial lighting, candlelight in particular, is more pronounced during winter. In this season, garlands of electrical lights decorate the balconies of apartments and mingle with the public street lighting. Nicole, a young student living in Islands Brygge, explains why she has strings of lights on the balcony all winter long:

> From sunset and all night long, until early next morning, because I find it so cosy (*hyggeligt*), when a little bit of light reaches the inside from out there. In the same way, I love to see the light on the other side of the street and see that people are at home. In this way, there is a little bit of life around me, and I can see that I am not the only one awake around midnight. Light offers a sense of secureness (*tryghed*) and community (*fællesskab*).

Nicole knows the curtain design and flowers in the windows of the neighbouring apartments, but not the people. And yet when people on the other side of the street light two candles in the windows, she does the same and feels, in this way, she is part of a larger community beyond the confines of her apartment (see Figure 5.1).

The notion of the neighbourhood and community is in this sense not about known people or a fixed spatial relation, but about a performative atmospheric co-production. As Jiménez and Estalella found, 'The neighbor is not a stranger, nor a friend, nor kin but a form of sociality whose value is an effect of ambience-experimentation. The neighbor is an atmospheric person' (2013: 121). Through Nicole's lighting practices of negative boundary-making, she feels that she helps bring the inhabitants together as a caring neighbourhood by collectively creating atmosphere in public spaces: an atmospheric community. Light, it is repeatedly noted among my informants upon reflection, is about care and caring. This may not be the only reason why she lights candles and electric lights, but it is a good bonus, as she explained. Her own personal presence in this sense extends beyond the walls of the apartment, and the neighbours are, in turn, present, but not necessarily visible in her life. This intimacy makes her feel that they have something in common, despite (or perhaps because of) the fact that she knows none of them – not even what they look like.

These sentiments were shared among many, but not all, of my informants in Islands Brygge, who would either not pull the curtains, or only pull them to the extent that they could still sense the light from the outside coming in. They would light their garlands on the balcony in order to 'move *hyggen* [the cosiness] out into the cold, in the public spaces around Christmas time', as another informant stated. It emphasizes, as lighting designer Roger Narboni's notes, that 'the very essence of working with light is the relationship with people, with what they feel' (quoted in Laganier & Pol 2011: 337). Yet it also highlights that it is not about individual feelings, but about how people come together around atmospheres and how a particular light comes to stand as an identity marker. It points to the shared norms and expectations of a neighbourhood. This chapter is therefore about how 'light offers a sense of

FIGURE 5.1 *Candles in the windows in Islands Brygge. Photo by Matilde Lykke.*

secureness and community', to use Nicole's words, by taking part in boundary management and shaping a sensuous community. It is explicitly about more *shared* notions of community, while the next chapter focuses on diversity in the meanings of light, juxtaposing the norms of sensing dimmed lighting in this chapter and the bright light in the next. The point is to show how this particular notion of community is concerned with sensuous performativity, rather than representation and fixed identities.

As covered in the previous chapter, atmospheres are vague half-things emerging in the co-presence of subjects and objects as subjective facts by being 'spatial carriers of attunement' (Böhme 1995: 22, 29, 1998: 114). They are the medium through which perception, and hence human action and understanding, take place (Böhme 1995: 48). It was also discussed that there is a distinction between *experienced* and *reflective* atmospheres (Thibaud 2014) – between the felt and the thought – and between the *felt* and the *staged* (Bille et al. 2015; Böhme 2013a) – the atmosphere and the atmospheric. Yet if atmospheres are 'subjective facts', then it raises the question of whether one can *share* an atmosphere. People do not simply become immersed in atmospheres in a void but are attuned through a culturally informed sensorium by the norms of what to expect of an event, by their upbringing and class, and by previous events. Temporality, anticipation and culturally informed sensoria are thus as central to understanding the everyday life of atmospheres.

These conceptual discussions from the previous chapters are employed in this chapter by looking at how the boundaries between private and public spaces are permeated or enforced through lighting. The routinized everyday practices of orchestrating and changing the intensity and quality of atmospheres are rarely emphasized in the expanding literature on atmospheres (but see Daniels 2015; Edensor 2012, 2015b; Hasse 2012; Pink & Mackley 2016). How does a particular atmosphere come into being, change, and become shaped in everyday life? How do atmospheres unfold as a co-creation between various people and things? In this chapter I discuss how light is more than a medium for subjective perceptions; it is a way of sharing and co-creating atmospheric spaces by manipulating boundary zones. The chapter shows how orchestrating lighting also creates, shapes and expresses people's attunement and ideas of domesticity, self and neighbourhood: how light is used to shape emotions *in* and *of* the home by staging the limits of its extension. Beyond consuming energy or offering visibility, lighting in this way connects to social issues of constituting well-being, community affiliation and care through enforcing or permeating boundary zones – windows in particular (cf. Helliwell 1996). The chapter moves from shaping individual spaces over more collective atmospheric practices, to a broader discourse about a particular Nordic light in the final part, all of which constitute a particular role for light in shaping collective identities of people and their things. Before embarking on such a journey into everyday lighting practices at home, let us first turn to the neighbourhood in question: Islands Brygge.

Islands Brygge: a brief introduction

Islands Brygge recently undertook extensive urban renewal, and a wide variety of architectural styles and people live in one of the most densely populated areas of Denmark. With around 12,000 inhabitants in a one-square-kilometre area, it is home to a very diverse range of residents: students, CEOs, artists, celebrities, sportsmen, academics, the working class, the unemployed and the socially vulnerable. They live in a mixture of apartments dating from the beginning of the twentieth century and more modern residential architecture built in recent decades.

The area was created in the late nineteenth century by reclaiming land, and it was home to large-scale industry, military barracks and residential areas. With the decline of the industries, and the relocation of the military barracks, large areas of Islands Brygge were abandoned and the industrial buildings slowly dilapidated from the 1980s onwards. By the middle of the 1990s an extensive redevelopment plan was taking shape, transforming the industrial area into a residential one and, in certain cases, reusing industrial components such as silos as a structural part of the new architecture, linking the past and present (Paddison 1995). Prices sky-rocketed, cafés, fashion boutiques and upmarket organic shops flourished, but large-scale gentrification has been somewhat mitigated, and the traditional so-called 'brown pubs' still attract regular, working class customers. The industrial docks have now been converted into the popular Harbour Baths, known as 'Copencabana'. The rapid transformation of the area has also resulted in a new nickname in the media: 'The Manhattan of Copenhagen'. Although Islands Brygge is part of the 2300 Copenhagen S postcode, which covers a large section of the island of Amager, where Islands Brygge is located, another unofficial postcode, 2301, has been established locally in an attempt to distinguish this part of Copenhagen from the other part, and its people, in 2300. Partly used ironically, partly as a commercial branding, 2301 highlights an idea of a particular Islands Brygge neighbourhood.

There are three architectural types of interest here: first, the late twentieth-century modern apartments generally inhabited by the middle and upper classes; second, the traditional early twentieth-century apartments, often home to residents from the lower or middle classes, but also the upper-middle class, who refurbished them or combined two apartments into one; and third, the barracks in Ballonparken, refurbished into a sort of independent commune. As a consequence, we see a variety of people and classes, also apparent in the way homes are decorated: from students' IKEA and inherited furniture, to the upper-class minimalism with Poul Henningsen *Koglen* lamps (costing around £5,000) and mid-twentieth-century iconic Danish design. The new modern apartments are characterized by large floor-to-ceiling window sections, often with 'conversation kitchens' (kitchen and living room in one) and large balconies, and mostly opening on to green areas between the buildings. The traditional apartments mostly have separate kitchens, small balconies and smaller rooms (and toilets) than the modern ones. Both the modern and traditional apartments are available as leasehold, commonhold and freehold apartments. The barracks in the area called Ballonparken are the oldest type of building. Named after the manned balloons used to spot the points of impact of the artillery, it was the training area until the mid-twentieth century for soldiers testing artillery canons. The barracks were built in the late nineteenth century, and have since been taken over and refurbished as an independent commune. A high degree of DIY is seen here, and there is only one electricity meter, shared among the inhabitants.

Connecting people with things

There is no such thing as daylight. At least not in singular. The sun moves during the day, during summer and winter, casting shadows from direct and indirect light, and clouds blot out the sunlight, leaving natural lighting levels hard to control. Even the way in which the human eye registers light changes (Boyce 2014: Ch. 2). In the recent architectural style of large window façades in Denmark, particularly evident in the modern apartments, interior light levels rise and lower rapidly according to the sun and clouds. Rather than accepting the changing light this has meant that electrical lighting is used as a stabilizer and to limit visual contrasts (Bille & Volf 2011). Such changing lightscapes impact on the feeling of a space, from the depressing grey of rainy days to the uplifting first gleams of a low-hanging spring sun.

Many of my informants arrange their rooms according to the daylight (B Hauge 2013a; cf. Hauge 2015). Naturally, the architect had already done much to predetermine what the inhabitants could do. It was sort of obvious from the floor plan and electrical sockets where the television, dinner table, etc. could be. Nonetheless, several informants mentioned how they had positioned a sofa in a particular place so they could see their favourite place on the skyline from the window; how the large façade windows 'open up the room'; how they were part of the outdoors; setting up the dinner table in a place where the morning sun could slightly 'touch' it; a place to enjoy dusk – also called 'the blue hour' (*den blå time*); or putting particular objects with a reflecting surface quality in places where the sun would move across to create a glow (*skær*) in the room. With the exception of blackout curtains in the bedroom, people rarely explicitly thought about light, artificial or natural, in everyday life. But when asked, many opened up and mentioned how they liked the light reflecting on certain things – or how the television or computer was not used in a particular place owing to such reflection. Natural light then shapes a feeling of things and spaces, attuning oneself and the atmosphere in a room.

Daylight, however, is not the only source of light in the homes. Even in the daytime, when most parts of the apartment are illuminated by direct or indirect daylight, many informants would often have a lamp turned on to 'avoid the home feeling bleak' as they say; meaning that the light entering from the window shapes variations of contrasts and shadows that are mitigated by applying electrical light in some areas. And even small things like diodes on televisions or electrical clocks emit a bit of artificial light. Throughout the day, electrical light is continuously being turned on, off, or adjusted, depending on the activity and atmosphere being sought; from a lot of light in the kitchen and toilet, to less in the bedroom (cf. Stidsen et al. 2011).

Catherine, a 30-year-old administrative officer, was particularly aware of how the light in her home was not static. She noted how the lamp above the dinner table that offers a lot of light is most often dimmed when no one is sitting there, rather than being turned off, since 'it gives a very good *hygge* in the room'. When she turns on the television she also dims the lamps that she used earlier for reading. She does not turn them off, just dims them: they are too bright to allow her to concentrate on the television, but it would be too shadowy without them on at all. It is about shaping an atmospheric interiority, most often in the peripheral vision. She and her husband also dim the light really low late in the evening just before bed time, instead of moving directly from television to bed, after a brief toilet stop. As they say, 'just to get that feeling of calm, and slow-down in tempo; creating a dusk-like atmosphere. It could be done with music, we just do it with light.' In other words, they continuously change the light in their home to shape the feel and atmosphere: a low light level to calm down, a high light level in a small spot when reading, etc. with the aim of attuning oneself according to activity. It is a practice not only based on putting light on the activity, but also a spatial focus on the visual comfort of the surroundings.

Furnishing the home with light

Light can be used to connect sections of the home and to attune the inhabitants to the atmosphere and activities. Elina, a high school teacher who often works in the evening said: 'Then of course I need light so I can see, thus turning the light bright. Then when I am done, I also think I separate work-light from spare-time light, by turning it down again. It [light] affects the division of everyday life.' Catherine, Elina and other informants illustrate a point: that reading a magazine and reading a work-related text are two different kinds of atmospheric practice – work reading and cosy reading – and by doing so also mark time.

Anne Bay, director of the Danish Lighting Centre, has noted that:

> our love for cosy light means that we do not always create the proper light for reading, sewing or cooking, activities that often demand a sharper and brighter light. Many people remember proper light in their hobby workshop, but forget it in the rest of their residence.[1]

The point from my informants above, however, is to nuance such notions of 'practice-of-reading' light, where light is shed *on* the text, and instead focus on 'atmospheric-practice-of-reading' light, where light is shed *for* the text. Light helps connect people to things – the magazine and the person for instance – and that atmospheric connection differs depending on what is being read or done. Of course, many people have poor lighting and are dissatisfied with the illumination they have, such as poor visual conditions in the toilet for putting on make-up. But people do not only light their room according to *what* is being done but also according to *how* that activity should feel. Particularly in terms of lighting standards, activities are held as the baseline for how to light a space – work light, reading light, sewing light. However, from the above we also learn that people use light at home to attune spaces and people. Thus, in the same way as music is often used to create an atmosphere, orchestrating light also sets the tone of a space for practices to take place in certain ways. The point is to shift focus away from how the body best accomplishes a task – reading, working, sewing – to a spatial focus, of how a space should feel while doing it.

To achieve such varied lightscapes, people often have between six and eight lamps in their living rooms, aside from candlelight: lamps above the dinner table or sofa, one general light source, for instance an uplight, standard lamp, or a ceiling lamp, and a reading lamp and table lamps in corners. As Bay also states,

> lighting is part of the home's furniture, and it can be used actively when furnishing. A room must have a good general light, and if you also have many lamps and use dimmers, you get the possibility to change the light so you can create different atmospheres.[2]

In line with Bay, one informant living in a single-bedroom apartment expressed how she uses many different lamps to 'furnish' (*møblere*) the room in different ways through light. She noted that one type of lightscape creates a workspace, another lightscape with less light creates an 'after work atmosphere', and when there is only candlelight it creates a 'spiritual space' (the informant worked as a shaman and alternative healer).

There are of course great variations in what is done in homes and to what extent light is used. People may not turn on the light in the bathroom or hallway but instead borrow light from other rooms. Some of the younger students and people who had just moved into a place had fewer lamps, but it was nevertheless often stated how light was used to guide different atmospheres, mostly under the headline of *hygge* even if it came in different shapes. For instance, Martin, a humanities student in his mid-twenties, was irritated that a particular lamp was not working at the moment:

> I really like when it is working. Then I can sit with my records and books and feel 'old' [*tussegammel*, which in this case has nostalgic connotations]. It throws a light down, so it is like there is only *that* corner. And I turn off all other light. That's my reading cave [*læsehule*] right there. I like that. In my grandfather's chair.

Clearly, light is not the only element in attuning him – which also included the chair, his records, etc. – but the practice of reading has become about much more than simply getting the right reading light. It is a nostalgic connection through the spatial and affective qualities of a light beam and the surrounding darkness.

There is also a general distinction between 'practical' (also called 'functional') light and 'cosy light' (*hyggelys*) among informants. Cosy light mostly entails the dimmed, dispersed light creating spaces within spaces, while practical light is often found in the kitchen and hallway, 'where you need to see what you are doing'. As Martin further explains,

> [w]hen you are eating you need a lot of light. Later when you are sitting and drinking coffee, I would then turn off some of the light. It [light] is in a sort of way part of creating the atmosphere surrounding what one is doing.

This notion of 'practical light', as should be clear by now, rarely entails a fluorescent light or bare light bulb completely illuminating a space, although some lamp trends, such as the popularity of the bare Muuto E27 lamp or the bulbs on the cover of this book, may promote glare (but then people have a dimmer for those lamps). Rather, spotlights may offer the general lighting or specifically illuminated surfaces, such as around the cooking area, while lampshades are set around other lamps in the kitchen. But just because a light is practical does not mean it cannot be cosy. As Martin found, 'doing the cooking may automatically become cosy. So, it goes hand in hand, even if practical lighting is the primary objective. But it may easily also be cosy'. As written in the pamphlet *Godt lys i Boligen* ('Good domestic lighting') by Lysteknisk Selskab, '[t]he kitchen is the most important work spot in the home [. . .] Therefore both daylight and artificial light need to be able to create *hygge* and simultaneously be effective, so you get good working light and avoid accidents' (2004: 6). What is central here is to see how light is continuously turned on, off, or dimmed either to support activities or to create a particular atmosphere, and, as illustrated in Chapter 2, Whilhite et al. (1996) remind us that such lighting norms are not universal or ahistorical.

'In many ways, light is something you do.' Eric seemed a bit surprised with himself as he realized that he had just made an analytical point about his lighting practices that was counter-intuitive to the taken-for-granted notion that light is a physical and measurable phenomenon. Eric's insight in many ways epitomizes the point in this book of seeing lighting, both natural and artificial, as a continuous practice of attuning atmospheres; of not settling with measuring the amount of light, but seeing how it came into being and why. Two issues emerge from seeing lighting as a practice: routines and care. Firstly, lighting is heavily routinized, and shapes ways of moving through the home, connecting person, place and practices (cf. Hauge 2015; Pink & Mackley 2016). Elina noted this when reflecting on her movements through her apartment. She remarked that 'darkness demands an action; that is, finding the switch, turning it on, and then moving into the room'. Knowing the position of the switches, and continuously turning on or off light, is part of home-making through atmospheric orchestration. Secondly, light is about caring. Nicole cared for the atmospheric neighbourhood, but it is particularly clear in households with more residents. For instance, in families with children, light may be turned on in the hallway when the children are put to bed to make them sense that they are not alone, or when the husband in one couple for instance turns on a light globe when he goes to bed if the partner is returning home late. Lighting is entangled in practices of caring.

Gathering the room

Electrical light, however, has some disadvantages. The position of the electrical fixtures may limit the ability of the inhabitants to position the light sources as they wish. One informant in particular, a student renting a room with only one electrical fixture, hated to have cords running across the floor, and as a result he had only a few lamps, mostly uplights, offering indirect light reflecting on the white walls. But it offers a static light both in terms of position and light cast. Unlike electrical light, candles are easily moved and can be placed on tables, shelves or windowsills. Very few had LED facsimile candlelight at the time of fieldwork, even if it may offer the same visual experience. There are three different attitudes to the LED candlelight. Firstly, those who use LED candles as a substitute for candlelight, because of concerns over particle pollution. Secondly, those who have LED candles as backup, in places where it may be more convenient or one would not notice the difference (outdoor, or in remote corners of the room), and thirdly, those who would never use LED candles as they are seen as kitsch and inauthentic. I relate the latter appreciation of candles to the task of lighting them and moving them into the right place. One informant, who consistently connected the flame with the term 'focus', explained how the very etymology of the word 'focus' comes from the Latin for 'hearth' and figuratively 'home'. She expressed how the act of lighting candles is 'like offering a cup of tea to a guest: a way of welcoming'. Central to the practice of lighting candles and dimming light is that people do it to remind themselves and others that they need to relax, that they are not at work anymore. This was shown at the beginning of the chapter where people came home from work and turned on a lot of lights, and as was shown with Catherine dimming the light before going to bed. Such affective properties of cosy light as dimmed light or candlelight even encouraged my own dentist to have a lit candle in the window of the examination room in the middle of the day, in order to take away the medical, clinical atmosphere that may unsettle the patients. Dimmed light transforms a boring or tense atmosphere into a (supposedly) more relaxed one.

Hence cosy light *(hyggelys)* both signifies and allows for the body to relax. Martin, for instance, often sits on the floor where he creates 'a tiny world' for himself. He can reach the books around him, reach the hi-fi and press play, and cuddle up with a duvet.

> Those are the kind of small universes you can create, which to me is *hygge*. If you can create a tiny, tiny world for yourself and isolate yourself and say, 'Now it is just me, and I don't know, a book or something, right?' So, you light up *this* area, and this is where *I* am. And it is great if there is sort of a small circle of darkness or at least darker illumination. Then you feel that 'now it is all about this point', and this is where we are concentrated. Shutting everything else out. Then nothing can stress or irritate or disrupt you [. . .] The surrounding darkness is all of the sudden there, embracing you with, like, a cosy layer, duvet, or cover. An embrace – a big lovely hug. I think that is *hyggeligt*. Darkness can be bloody *hyggeligt*.

This emphasizes how, through lighting practices, the manipulation of boundaries, visibility, cover and embrace, helps orchestrate the particular kind of atmosphere sought. *Hygge* both reflects and constitutes a particular kind of quality in life: people feel good, feel safe, in the co-presence (*samvær*, literally together-being) or intimate presence (*nærvær*, literally near-being) of other people, or just by themselves with their selected things. Illumination becomes a gathering point, and shapes the atmosphere where my informants feel at home. The home may consequently be staged with all the right things to make it 'atmospheric', but actually feeling at home is about attuning oneself.

Embracing ecstasies

Everyday practices such as vacuuming, cooking, reading, working, watching television are often imbued in efforts to attune oneself to them. As one informant explained, she likes to read in the bedroom where there is a comfortable armchair, and a lamp, rather than in the living room:

> I really like to sit there because, if I have the light on in the living room, I can see the glow [*skær*] through the glass door. It creates this effect of an extended room, and that there is a warmth in the room, in some way.

Others such as Catherine, having trouble making the modern apartment fit her idea of cosiness, tried to 'pull the kitchen into the living area' in her kitchen-cum-living room by putting a particular cosy lamp in a place, so that the kitchen no longer had what she calls a 'kitchen atmosphere' (*køkkenstemning*). The lamp 'gives' *hygge*. In this sense, shadows and patterns from lamps take part in 'embracing the room' (*omfavne rummet*), as another informant noted. Lamps not only illuminate but leave a mark.

The above example of appreciation of the quality of luminosity is what Böhme terms *lightness*. He notes that the term 'indicate that it [the light] does not have the character of a thing but of a freely floating quality or [. . .] a quality that flows indeterminably into the vastness' (2017: 198). More than simply an observation on perception, it is a matter of attunement. One informant, Anna, spoke keenly about the way she orchestrates lightness to create an atmosphere when home alone:

> I use light to create a presence beyond my own. It is not something that I have previously been so conscious about. I have been studying for exams and my boyfriend has been away for some time now. I've been sitting in the office, writing, sitting there, and still I just kept the light on in the living room. It is not an excessive light; it is not glowing vividly, just a little bit. Enough for me to feel that I am not alone. That's how I feel. Just a little bit of light in the living room makes me feel like something else is there [. . .] It is a presence that is not human. And it is not sound, although I think I also use sound sometimes, turning on the radio [. . .] But yes, I think I feel that

> there is someone other than just me, if I put this light on. I can't put my finger on what it is.

Anna, like Nicole in the beginning, uses light to attune herself by staging an affective presence, even if at a distance. It is not so much about feeling lonely as about staging a particular wished-for solitude. People use sound, television or light to shape a feeling reminiscent of 'having company', as one informant stated. Anna continued expressing how light has a particular material character, in that:

> light reaches farther. A table has its own fixed borders: here it is. But light is something you can sense much farther away. So, if I turn on a light in here, I can sit in the room next door and sense that it is there.

This reminds us of Junichiro Tanizaki's insight that 'we find beauty not in the thing itself but in the patterns of the shadows, the light and darkness, that one thing against another creates' (2001: 46).

A philosophical question about the beginning and end of a thing is raised by such observations of light. Through the quality of colour reproduction and temperature, the light source affects the way surfaces, and hence objects, appear. Böhme calls this the 'ecstasy of things', connoting an object's ability to step out of its own tangible boundaries (*Aus-sich-Heraustreten*) (2001a: 131) and impose itself on the environment. He uses the example of how a blue cup also casts a blue shadow on to other objects, whereby a thing articulates its presence through 'tincturing' other objects (1995: 32–3). A tea light on the corner table may, in this sense, connect the immediate surroundings with other parts of the room by casting shadows, or the lampshade may change the glow and shadows from a bulb in the living room, shaping the feeling of reading in the bedroom. This perspective on ecstasies highlights Heidegger's insight that 'a boundary is not that at which something stops but [. . .] the boundary is that from which something *begins its presencing*' (1971: 154, original emphasis). Objects extend into space, among other things, by the light they are embedded in and embraced by.

More than a 'symbol' or 'sign' of something else, light shapes a presence in the world that ties closely into notions of self, solitude and comfort. It is a matter of 'bringing life' into the home by connecting each room. To my informants above, the power of light is clearly not a matter of optimal distribution or quantity of light for the sake of visibility but about a very embracing and affective lightness creating an atmosphere by connecting things and places. It is a dual presence that makes things visually present (or absent) and through this provides an affective presence and embrace of a home that ties into notions of comfort, care, and secureness.

Offering glow

From listening to informants and being in (and outside) their homes, it is evident that lamps not only illuminate but also leave a 'gleam' or 'glow' (*skær*) in the room and on people. Light does not neutrally reveal what is there but impacts on the visual presence of the material world. By arguing that light shapes a particular presence *in* and *of* the world, I do not intend to create a binary opposition between light and dark, visibility and invisibility. Rather, I want to emphasize how the interplay between intensities of lightness, darkness, glow and shadow becomes an element of the continuous modulation of atmospheres through the informants' lighting practices. These atmospheres may not be noticed or may be taken for granted, but nonetheless they hold an intensity that is orchestrated through material culture, including light. Elina was one of the informants who spoke in most detail about the role of glow as the quality of lightness, and the ecstasy of things:

> The 'living light' [candlelight] is particularly capable of catching and offering reflection [*genskær*, literally re-glow] in glass beads and glossy surfaces [. . .] Glow, I guess, is the – I was about to say product of light – but what is cast from the light source [. . .] It is probably the glow you register unconsciously, because it, like, implants itself [. . . T]he glow from the coloured glass beads against the wall gives such a *hyggelig* – I would almost say joyous – effect. It is the glow that creates the atmosphere – an attuner [*stemningssætter*]. And it creates life, you could say.

Light puts a mark on what is to be seen or glimpsed, and emphasizes the darker spots (Figure 5.2). As one informant, Richard, who had comparatively darker rooms than others, noted, a room should not be so illuminated that the darkness disappears: 'Light should be part of the darkness.' Darkened areas slipping into illuminated areas are what create my informants' cosy atmospheres.

When lighting specialists in Denmark state that 'light is life', this may be a biological fact but it is in the half-light that people live – whether alone or with others (cf. Borish 1991: 269). The use of cosy light has reached the point where it has been claimed that the flame, the living light, is at the heart of Danish domesticity (Ardener 1992: 28). While this holds some truth, it more often seems to be the glow, the shadows and the darker spots of the subdued lighting that direct domesticity, blurring or even visually dissolving the gaps between things and spaces, allowing people to sense the home 'from the inside' as habituated and embodied, rather than as separate from the person. In this sense, the light from the bulb, candle, sun or moon may visually present the world, but it is in the nuances of lightness, darkness and shadows – in the

FIGURE 5.2 *Carving out spaces of light and darkness. Photo by Matilde Lykke.*

absences and invisibilities forged by the ecstasies of the light sources – that the particular atmosphere comes to life. It does this by oscillating between connecting and separating the boundaries between things and people. Yet there is more to lighting than creating intimate relationships between individual persons, places and things.

Being together through light

Nicole uses light to connect people and transgress boundaries between indoor and outdoor for the sake of the neighbourhood. Like others, Nicole explicitly expressed a sense of community feeling with the people on the other side of the street when their lights were on. This may happen in everyday life, or it may happen on more celebratory occasions such as 4th of May where some people light candles in the windows as an act of remembrance of World War II liberation. Most informants in the traditional apartments did not feel visually exposed, since very little can generally be seen through the windows, with plants taking up some of the view and curtains often being pulled (cf. Garvey 2005; van der Horst & Messing 2006; Petty 2012). Moreover, the subdued lighting makes people less visible from the outside, offering only a silhouette, consequently shaping some sort of privacy in itself.

When observing the modern apartments with larger glass façades in the evening, the situation is different, as the inhabitants are more visible. With some exceptions[3] most inhabitants are not bothered by the increased potential for visual exposure that illuminating the home offers. Although many informants draw their curtains in the evening so that no one can gaze in, more often this is linked to the activities inside than to a sense of being exposed. The informants stated that, when alone, they might draw the curtains in the evening to create a small cave, while if they have guests they do not bother about other people gazing in, since it also extends the boundaries of the room to be able to see the lights in the city. In the city, the windows do not turn into a black surface in the evening mirroring the inside due to the lights from the surrounding apartments permeating the windows. In the new apartments, informants even said they do not draw their curtains in the evening until they go to bed because their visual exposure – shaped through illumination – signals a kind of openness to the other people in the area. In other words, shedding light and allowing visibility are ways of connecting with people and creating an image of openness. As more people share this *will for visibility*, the more the neighbourhood comes to express such ideals.

Distance and nearness

Informants such as Nicole highlighted how there is a sense of community shaped through visibility and lighting practices, yet it is a community in which they may see their neighbours but do not know them; that is, as a person from whom atmosphere radiates through their practices, often at the margins of attention. They are kept at a distance, albeit close enough to create a sense of community. This is also the case even if curtains are drawn, as there is still an ambient light emitting from the windows.

An example of overstepping this boundary came in an interview in the modern apartments where a woman explained how she would stand in her home and wave to people with large glass windows on the other side of the street but would never talk to them on the street if she met them. To the partner, this was overstepping the line of privacy, in the same way as when guests, unaccustomed to façade windows, sometimes stand and stare out of the windows into the other apartments. Waving reaches a point where this presence-at-a-distance is threatened. There is an element of knowing who the neighbours are but, at the same time, a sense of privacy that stops short of actually engaging with them. Light and movement in the apartments implicitly connect people at the margins of attention, but direct communication is a more delicate matter.

Another informant, Margaret, was caught in a paradox in this regard: she did not like the fact that people could look into her modern apartment. On the other hand, if the apartment was completely sealed off in the evening, she felt her home had an uncanny sense of loneliness (rather than solitude) to it. Most often she would therefore have the vertical slats in front of the window open enough for her to sense the outside while limiting the ability to fully expose the interior, particularly when the lights were on. Other informants also expressed how 'cosy' it was to look out of the window and see the same people exiting onto the streets every morning, creating a kind of 'neighbourhood *hygge*'. On the one hand this again emphasizes the vagueness and plurality of atmospheres that *hygge* denotes. It is a reflective atmosphere entitled *hygge*, but that may differ from experienced atmosphere. But it also shows how the lighting practices are normalized in this area, with a large degree of homogeneity, even if not all inhabitants share the sensuous attention to their neighbour or wish to emulate the practices. People are more or less sensing in the same way; few stand out by having radically different lighting. It is a shared atmospheric experience, because the spectrum of lighting differences is narrow. The atmosphere here is thus also a product of homogeneity, while in other places, with more diverse lighting practices, that diversity may either be appreciated as the characteristic atmosphere, or considered alienating as it highlights that people do not share aesthetic values.

The concern about exposure is countered by the issue of welcoming the outside. Returning to Nicole again in the traditional buildings, she explained how light was like the sound of one's neighbours as they cross the floor or play music. It is a confirmation that you are not alone: 'Oh, I'm not the only one awake at one o'clock,' she commented when talking about the light on the other side of the street. Nicole liked the feeling that people in the neighbourhood had something in common, even if it was only the sensory experience of light from two candles in the window. It was 'cosy' and made her feel less lonely. Natalie, a resident of one of the modern buildings, had recently had a baby and complained that she could no longer go out to cafés or be a part of the city as much as she wanted. From her modern apartment, however, she and her child could sit and watch the city 'entering the apartment', as she put it. The ambulances in the street, the street lights, the offices and the light from Tivoli amusement park in the distance, watching how people moved around in their apartments became like a television programme that brought the city, its people and lights, and the interior together; distanced, yet shared.

Natalie would also rarely draw her curtains, since it made her feel comfortable to know that there was life around her. By lighting up their home, the lighting practices of anonymous inhabitants become converted into a norm-governed sensuous community. There is a profound sensory spillage on to the neighbourhood of what essentially are domestic practices of shedding

FIGURE 5.3 *Garlands extending boundaries. Photo by the author.*

light. The disruption of such norms becomes evident when, for instance, excessive strip light is commented upon in remarks such as 'that looks like an immigrant home', or the opposite, when no lights raises suspicion, or when a dark area appears desolate and lifeless. The point between distance and shared is thus important because the positive notion of being part of a greater community illustrated by Natalie and Nicole, would also have it limits, as it is apparent that it is not an openness to everyone. It is about an openness towards the familiar atmospheric person whose practices resemble one's own, but not necessarily the stranger, or the one who does not share aesthetic practices or values.

Light, and the visibility it offers, hence becomes a means of caring and connecting with, or separating from, other people, and with a social life of which they may feel deprived. Light from the outside is just as important as light from the inside in creating an atmosphere. This was also the case in instances where people had candles in a lantern on the balcony outside. The large windows create a sense of being both inside and outside at the same time, and the balconies – often running the full length of the windows – are therefore used to extend the interior by placing lanterns or light garlands on them during the winter. The window becomes a way of extending not only the apartment but also oneself as a person, as a part of the city, where the external and indoor lights connect, comfort and contribute to the neighbourhood atmosphere, albeit also offering distance.

People are very aware that watching, surveying and staring are separate ways of seeing. The issue of bringing in the city and connecting with people is hence not about visibility, as such, but about the quality of seeing, where 'the public and private spaces are folded visually into, and shaped by, each other' (Stender 2006: 36). Nicole further illustrates this with her sense of a nested community shaped through connecting people and things, while also keeping a distance:

> It is what I love about living in an apartment, and I am not sure if I could do without it: the fact that I can hear people moving around. It is the same kind of thing that light does to me. There is someone still up watching television or reading; that to me is the cosiest (*hyggelige*) thing in the world.

Safety or secureness

Hygge is intrinsically connected with the word *tryghed* (commonly translated as 'safety'), which several scholars feel epitomizes the essence of Danishness and domesticity (Linnet 2011; Schwartz 1989). Rather than translating *tryghed* as safety or security, I propose secureness, which does not evoke the same sense of safe home in opposition to the threat and danger of the outside. This interpretation of *tryghed* as secureness is akin to the German notion of *Geborgenheit* as a feeling of sheltered-ness, nested-ness and well-being (cf. Bollnow 1963: 129–32; Hutta 2009). This sense of shaping social closeness through light relates to a central feature of *hygge*; that of feeling relaxed and nested. Judith Hansen describes this as follows:

> The feeling of protectedness within hygge's embrace tends to carry with it a sense of being provided for, of having one's basic creature comforts satisfied. Both these features thus contribute to relaxation and commitment to the event itself, as well as to the sense of social, physical, temporal, and experiential boundedness associated with it.
>
> HANSEN 1976: 57

By stressing secureness as nested-ness, I emphasize the way in which spaces are shaped through light to evoke intimacy and relaxation, not as the opposites of threat and danger but as a particular kind of comfortable atmosphere.

Collective lighting

To further illustrate the point about a sense of community being shaped through atmospheric lighting, let us turn to Ballonparken. There is only one collective electricity meter in Ballonparken. With its roots as an independent commune, there is a particular sense of collective identity in this area. Yet, if there is only one meter, the cost of electricity is also shared, leading to negotiations of the propriety of spending energy, for instance on lighting. Some people have larger houses than others, while others use more electricity because they let computers or lighting stay on even when away or they fall asleep.

As many people are continuously refurbishing their houses, work light is often on, which uses a lot of wattage. Forgetting to turn lights off is not just an economic but a moral issue. To one informant, wasting energy on lighting showed that some inhabitants did not share what she felt was the spirit of solidarity and community. In essence work light shapes an atmosphere of expectation – that people are actually doing something – as opposed to when they are using cosy light, which does not carry the same moral demand for efficiency. But sometimes the balance is difficult. Years ago, some houses had garlands of lights on their houses for Christmas because it looks beautiful and creates a festive atmosphere. To others it was light pollution cutting into the relatively darker night in Ballonparken, located close to the green area of Amager Fælled. Yet to others it was simply a waste of money. It ended after a few years, but it highlights the balance between atmosphere, economy and climate.

There is a strong sense of community in Ballonparken, and light helps to shape it (although light is by no means the only aspect). When a fuse blows, inhabitants meet up in the darkness and try to figure out what caused it. One informant compares it with a treasure hunt, and finds it really cosy. The absence of light becomes a way of shaping social situations and collective action. It reminds her of the dream of a primitive and simple life. Informants like the way everything in this sense is connected in Ballonparken: the kettle, the lighting, the computer, etc. Several informants also mentioned that they do not like it if people feel outside of the community. For this reason, one informant does not draw her curtains all the way; just enough for her to feel privacy, but not so much as to seclude others from sensing her presence. One may also find an outdoor campfire, but 'you have little chance of making a fire and sit alone. There will always be someone who comes over and starts chatting. It is a bit of *hygge* we gather around.'

Nordic lighting

So far, the chapter has moved from a sensuous orchestration of the home when alone, to the sensuous ways of shaping community. It is now time to explore a broader narrative of a particular light that is shared, not just at Islands

Brygge, or in Denmark, but across the Nordic countries – what is commonly described as *Nordic light* ('Nordisk lys' unlike Northern light, as the Aurora Borealis, 'nordlys'). My informant Elina reflected on whether the use of candlelight was a particularly Danish thing:

> It is definitely the idea that the farther north you get, you also get more dark hours during winter season. And thus, you turn more towards both heat and light sources that emit heat, so to say. So that is why it [candlelight] is more widespread in the Nordic countries. Probably also in Canada or Alaska. I wonder if they have it in Russia? It is also cold and dark during winter there. Maybe it is mostly in Scandinavia.

In Denmark there is a narrative of a particular kind of light in the Nordic countries because of the geographical location, aside from the promotion of a national trait of *hygge* as political soft power (Howel & Sundberg 2015). But, as Elina also highlights, the question then arises as to whether the particular light is caused by the geographical location or whether there is more at stake. Farther south in Europe, the Middle East and Africa there is, for instance, a preference for stronger direct white lights compared to dimmed and reddish glows farther north. This may be described as *geographical determinism* in which people's lighting practices mimic the course of the sun. The argument goes that the short transition to night around the Equator is mimicked by strong interior light, while long transitions farther north make room for more reddish glows.

The notion of a particular 'Nordic light' has become popular in Scandinavia, claiming that there is a particular natural light in this region that guides our lighting practices and preferences: 'The beautiful Nordic light has represented something special for human-beings since the very dawn of time' (Sørensen & Haug 2012: 1). At a seminar at the Royal Library in 2012, the theme was *Nordic lighting: How to make Nordic lighting design an international brand.* It tapped into an international narrative of Nordic cuisine, film, design and architecture, evoking a shared sense of cultural identity and aesthetics. The task at the seminar, with participants from across the lighting industry, was to discuss and develop the traditions of lighting that shaped Nordic architecture and design in the twentieth century.

Yet an interesting tension is also apparent in the promotional pamphlet: sometimes Nordic lighting as a cultural appreciation and tradition of working with light, and sometimes it is Nordic light, as a particular kind of light. In both cases the notion of Nordic light(ing) shapes an idea of cultural coherence through an atmospheric community that stops at the border to Germany, never crosses the Baltic Sea to Russia or the North Sea to Britain. Nordic light(ing) is what is sensed and a way of sensing. From the many excerpts from my interviews in this book, it is also not difficult to see that there is a particular appreciation of light offered by the northern hemisphere in Scandinavia, reflected in the (changing) aesthetics of dimmed lighting and orchestration of boundary zones. It is a cultural aesthetic that architects, designers and ordinary people have appreciated, developed and moulded. The emphasis on letting abundant daylight in on white walls, promoted as emblematic of Nordic light, however, needs to be set in a historical context. It may be *comme il faut* to have white walls now, but in the 1990s interior design trends promoted walls painted red or yellow with a sponge, and there were the 1970s predilections for brown. The appreciation of light and colour changes.

While I find it easy to accept the shared aesthetic appreciation (Pallasmaa 2011) emphasizing a collective sensorial norm, or a definition where it is above 60 degrees north, where the night never turns dark in the summer (Garnert 2011, 2016), there is, however, something more troubling about the notion that Nordic light is a particular light *only* visible in Scandinavia (Matusiak 2012). The difference is essentially between Nordic lighting as *practices* with light and Nordic light as an *essence* of light. Copenhagen, Glasgow and Moscow are approximately on the same longitude, but people have very different lighting practices. The counter-argument often goes that the particular climate in Scandinavia is different (e.g. Matusiak 2012; Norberg-Schulz 1993). Thus, while Scandinavian countries are

> dissimilar in topography and vegetation, their skies share a subdued light that imbues the entire region with mystery. More than the landscape, it is this dream-like atmosphere that tells people at once that they have reached the outermost rim of the earth.
>
> PLUMMER 2012: 6

This, however, downplays the enormous variation in both natural light and weather from the Aurora Borealis north of the polar circle, to the snow-filled fells of Sweden, over the fjords of Norway, to the hills in Denmark. Furthermore, at times the 'Nordic countries' also include Iceland and Finland, where one learns that '[g]rowing up in the special Nordic lighting conditions thus naturally affects Nordic people' (Sørensen & Haug 2012: 33). Yet this reified version of Nordic light, restricted to Nordic national boundaries, does not include people in Tallinn, Riga, or St. Petersburg, all similarly close to the Baltic Sea and positioned on a longitude between Copenhagen and Helsinki. There is a distinct cultural boundary-making process involved in the concept of Nordic light(ing). Another way of understanding a reified Nordic light is to follow Nanet Mathiasen (2015) who explored Nordic light in case studies solely located between 56°N and 62°N longitude. Although her cases were from

Scandinavian countries, the point here is that national borders are not the defining factor; the particular light is.

The argument here is that whether or not we are talking about Nordic light as a material essence, or Nordic lighting as a shared sensory appreciation and practice of lighting, they are both part of establishing a shared Nordic sensorium and identity. 'Nordic light' highlights a bounded community, sensing and living within a primordial essence of a particular light, while 'Nordic lighting' stresses a community of shared lighting practices. These imply the distinction noted in the introduction between a *climatic* atmospheric essence (the former) and *existential* atmospheric practices (the latter). It is such a particular appreciation of light that unfolds in the domestic practices of permeating and enforcing boundaries between people and things through light.

Atmospheric community

One way of understanding the sense of community illustrated with Ballonparken, Nicole who would light candles like the neighbour, discourse of Nordic lighting, or the application to have *hygge* as UNESCO intangible heritage, is to be inspired by Benedict Anderson's notion of *Imagined Communities* (1983), which denotes the sense of community shaped through cognitive and symbolic structures beyond social relatedness. Anderson argues that the nation 'is imagined because the members of even the smallest nation will never know most of their fellow-members, meet them, or even hear of them, yet in the minds of each lives the image of their communion' (1983: 6–7). This notion of imagined community allows us to look beyond social proximity as a defining factor, as also indicated by both Nicole on a small scale and the Cannon of Denmark on a larger scale.

Yet an imagined community remains a somewhat institutionalized, symbolic and cognitive construct, forming a 'we' in opposition to 'them'. Hence, while Anderson helps us understand how a sense of community may connect people who are not related through kinship or social ties, the notion of imagined community leaves us ill-equipped to understand the aesthetic sensibility connecting self and other as an atmospheric unit. That is, how does a community, which does not necessarily have any other essence than the normalized sensorium of an atmospheric neighbour, come together? It is a sense of community experienced at the margins of attention, but not defined by institutional or territorial structures. It is sensory performative, rather than about representation. There is not necessarily a collective 'we', or moral codex, yet there is a sense of togetherness – *fællesskab*.

In this sense, I am talking about a limited kind of community as it is based on collective sensory norms of *being-with* others, yet also unlimited as there are no clear borders. Thus, it is not quite a community in any traditional sense, but neither is it not a community, as the term (*fællesskab*) is indeed commonly invoked by informants. It is, of course, also tempting to see it as a community bound up on taste (cf. Bourdieu 1984), yet the point here is not the judgement of an aesthetic sphere of design, class, interests or properties (cf. Lash 1994: 157ff); community is a term clearly constructed at both political and discursive levels, but it is also tied to everyday practices. In particular, it relates to sensuous ways of orchestrating lightness whereby 'it is a communality of forms (ways of behaving) whose content (meaning) may vary considerably among its members' (Cohen 1985: 20). It is not a strong community in the sense of necessarily identifying with being from this or that area and not all residents may wish to contribute to it as explicitly as Nicole. In this sense, the atmospheric community may not even be based on a premise of difference from others or be essentialized with clear territorial borders. It is established by normalized and skilled ways of seeing the world at the margins of attention, such as sensing the sound, movement or light of the atmospheric neighbour.

Accompanying the sense of community described above is a moral moulding of what counts as correct behaviour and ways of sensing, for instance not staring through each other's curtain-drawn windows. Likewise, there may similarly be established other kinds of atmospheric communities in areas with more varied lightscapes. The point is the acceptance and identification with such sensory engagements and effects, not the homogeneity of the material expressions. The particular atmosphere is, in this sense, not something essentialized or fixed but is continuously established through practices; the community is shaped through a normalized experience of lightness. By approaching community from an atmospheric perspective, a sensory element is added to the dominant discursive and symbolic approaches.

Angharad Stephens argues, 'The concept of atmospheres leads us to think about how collectivities come together and disperse, about the spatial architectures that hold those collectivities together in time' (2015: 100–1). Viewed from this angle, atmospheres are the affective relationship or orientation shared by a collectivity of people (Großheim et al. 2014: 4–5). The above highlights what Sarah Pink also found, when noting how it is through practices 'that residents engage in – which produce identities and embodied sentiments – that the idea of the community as a field of persons who have the potential to be connected to each other can be realized' (2008: 179). The sense of community that is established in extension and beyond the atmospheric person is located 'in the sensory, embodied sociality that is integral to their place-making practices' (2008: 184). It was a sensuous based atmospheric community, which was not so much about the specificity of taste and design but about sensory imprints. There was no common goal, although a norm-based light practice existed in terms of *hyggelys*, and it did not delineate a clearly defined community, or metaphysical or moral aim. Thus the notion of community invoked here is different to those normally discussed and criticized in social science (Delanty 2009). It was more than an individual *being-in* space, but a *being-with* and the consciousness of an interpersonal connection through atmospheric practices, however ephemeral and impersonal it may be: a vague atmospheric community. It is a community sense that became part of establishing a sense of home and recognizing the role of the experiences at the margin of attention, that had the recognition of a norm-based lightness as the base. In sum, the sense of community illustrated among my informants in Copenhagen was of a sensory type; one that emphasized practices and lightness.

It is of course by no means unproblematic to invoke the concept of community to describe social collectivities. Community has been used to cover very diverse categories of sociality, from 'traditional' ways of life under threat from modernity, to the national or transnational collectivities, or the more vague 'social bonds', reaching the point where one may be better off abandoning the term altogether (cf. Amit & Rapport 2002; Castells 1997; Delanty 2009; Turner 1966). Yet from the discussion above, and from the way the Cannon of Denmark and informants like Nicole mention that light and *hygge* create a community, perhaps it is worthwhile to focus both on the community aspects of atmospheres and the identity processes they partake in. Beyond community as a discursive or narrative phenomenon, it could thus be worthwhile thinking about its atmospheric elements.

Conclusion

In this chapter, I have shown how domestic atmospheres, *hygge* in particular, are staged with care for visual comfort through glow, shadows and convoluted spaces, often opening up to the outside. Light is used to attune activities; it shapes one's attachment to the activity, whether it is something that should simply be dealt with quickly or something one can dwell upon – for instance work reading and cosy reading. Light does more than just shine on the world; it is continuously used to create physical orientation in a room and mentally to attune to the situation. It is also a dynamic part of shaping the intensity of the atmospheres into what they are or should be: the atmosphere and the atmospheric. By intensity I do not imply that they are always noticed, are of the same type, or are easily discernible. Danish homes and lighting practices are about much more than *hygge*. Atmospheres emerge, even if unnoticed when coming home, taking one's shoes off and collapsing on the sofa, or they are noticed when stepping into a room lit by candlelight, with soft music and a bottle of wine, to meet your partner. In the homes of Islands Brygge, atmospheres are emotional orchestrations of space over time, where lights are dimmed, turned off, curtains drawn and candles replaced as time goes by, continuously negotiating the boundaries between the inside and the outside, and between people and things. Lighting is something people *do*.

One of the central features from the examples is the way light bulbs and candles transcend their own tangible borders and impose themselves onto the material world shaping a lightness. Lightness, and the atmospheres it co-shapes, extends beyond the borders of the apartments and shines its ecstatic nature onto the neighbourhood, bathing the surroundings in a gleam of *hygge*. Consequently, light shines to offer 'a sense of secureness (*tryghed*) and community (*fællesskab*)', to follow Nicole. In this way atmospheres become the medium for expressing and unfolding social and emotional life through the visibilities that light and darkness offer. This may happen through creating 'reading caves' of solitude, or through sharing the atmosphere with anonymous neighbours. A person may thus step into a radically new atmosphere shaped by things or co-created by neighbours. This partaking in atmospheres connects individuals that may otherwise not feel or seem connected by tying together the 'outside' urban spaces with the 'inside' domesticity.

Not all informants were of course as aware or interested in lighting or community as Nicole or Anna. But for these informants lighting practices create a sense of community among people who do not know each other, and who may not even want to for the very reason that knowing them would accentuate the differences that are often more pronounced than their lighting practices. This community approach to atmosphere and lighting as a practice of enforcing and permeating boundaries offers a more dynamic and socialized lens of understanding human embeddedness in the material world. Lighting practices and atmospheres, in this sense, may connect the individual and the neighbourhood, even to the level of the region or nation (e.g. Stephens 2015; Zerdy & Schweitzer 2016). This chapter has deliberately focused on shared lighting practices with little mention of conflict or alterity. *Hygge* is not to be treated as paradigmatic of atmospheric communities. Such communities may also be shaped in many other ways than through cosy light. The next chapter zooms in more on the diversity in what counts as 'proper lighting', offering a counterpoint to the somewhat homogeneity illustrated in this chapter.

6

Qualifying bright lighting

Introduction

The light from the incandescent light bulbs has been normalized during the twentieth century in Denmark to the point where other light qualities are seen as 'alternative' (Jensen 2014). Lamp designers, such as Poul Henningsen, and the inspiration his lamps have given to other designers, have taken part in normalising dimmed, glare-free light as quintessentially defining 'good lighting'. Spaces are supposed to appear in a certain tone of light. As shown in Chapter 2, this does not mean that what counts as visually good lighting has not changed historically. While light shapes how the world is perceived, there are also strong-held meanings and symbolism associated with it; a dark window or place may be seen as suspicious; red lighting connotes sexual activities; depending on culture, a candlelight may be associated with *hygge*, death or romance; and as of late in Denmark, fluorescent un-shaded bright light such as strip light in apartments is associated with immigrant homes.

Illustrating this latter point, the well-known weather reporter Peter Tanev in the morning show *Go' Morgen Danmark*, told the audience how he had tried the 'Wake up light' – an alarm clock lamp that slowly lights up, imitating the sunrise and triggering circadian rhythms.[1] He explained how his neighbour must have thought that a 'family with strip lights' had moved in. Unlike the somewhat homogenous dimmed lighting presented in the previous chapter, alternative luminous geographies are apparent in social housing areas with a high percentage of immigrant households. Large-scale immigration from the Balkans, the Middle East, Africa and Central Asia particularly increased in Denmark from the 1970s; with this also came new practices of lighting the living room and bedroom areas with strip lights, which had hitherto been designated to work spaces, attics or basements in Danish homes. In the same way that a large quantity of satellite discs on the balcony of apartments in a social housing is indicative of immigrant presence, so too is strip lighting.

The social housing of Urbanplanen, close to Islands Brygge on Amager (part of the same postcode), is one such multicultural area. One of the inhabitants is Janni, a Danish woman in her mid-60s in early retirement and who has lived most of her life here. During the interview, she is hesitant to generalize about lighting, but she has really noticed how immigrants just have strip lights on the ceiling: 'it does not matter to them what light it is, it is just some light,' she says. Janni is clearly putting a negative value judgement on what she calls 'bluish light' in the homes, but also accepts that it is their choice – even if it is not the *hyggelys* she is used to. Yet in her reflections upon the neighbours' lights, she also shows how the strip lights not only stand out visually, but also invoke both prejudice and presumption. As she stands on the balcony smoking in the evening, she mentions that she would sometimes contemplate about the people living on the other side of the street,

> Some have light on, all night long. And I have actually thought: I wonder if they have been victims of torture? I know it is a weird thought, but like, someone who is afraid of the darkness. There is always light on, always. It is a thought: I wonder if it is a victim of torture who has been sitting in a dark cell. It's just the thought I get.

To her, domestic lighting says something not only about the cultural origins of people living in an area, but also something about the meaning such lighting elicits. An apartment where the blinds are always drawn raises suspicion; the candlelight in the window, as seen earlier, indicates *hygge*; and bright light and lamp design tell us something about resident composition.

In this chapter I explore the social logic of bright light, and the continuous process of appropriating spaces through light to make a home. That is, pluralising the definition of 'good lighting' illustrated with *hygge* in the previous chapters. The key argument is that bright light shapes spaces in ways that make affective and social sense to the inhabitants. Lighting is an atmospheric competence and practice that is learned and changes according to biographical, social or political context. The narratives in this chapter are shaped around few individual households, the first four from Urbanplanen (out of eight interviews) and the last from Jordan. But the particular location is not the issue here. The point is not to say anything representative of light practices or home-making in Urbanplanen as an area with many first- or second-generation immigrants – the diversity in terms of family structure, cultural background, and income is an empirical fact, which is simply too great to find similarities through qualitative research. Rather, the point is to illustrate how lighting preferences are entangled in notions of home, privacy, and social gatherings and differences. It is to see how lighting preferences other than dimmed *hyggelys* make sense to individuals, and hence may guide us to better understand the social work lighting does, beyond a merely technical or physiological perspective.

Confronted with the Otherness of brightly lit homes forces researchers and people in general to rethink the taken-for-granted nature of 'good lighting' being either glare free, resembling the sun, or universal (for a nuanced discussion of bright light in Japan see Daniels 2015). This chapter is about qualifying the use of bright light in homes; about not making *hygge* the paradigmatic measure of lighting in Denmark. As with hygge, there are cultural logics embedded in lighting practice that may not live up to what is physiologically or psychologically optimal. The bright light may also be somewhat alienating, as seen with Janni, in the same way as *hygge* may have excluding properties by avoiding anything that may cause conflict or disagreement. Light, and architecture more generally, is offering material solutions, or challenges, to social norms, and with increasing immigration of people from distant countries, such diversity of solutions and challenges may increase. To illustrate how bright lighting unfolds and makes sense, I first turn to fieldwork conducted in Urbanplanen. This will be followed by another case from Jordan, intended to further highlight the multiplicity of meanings and practices of shaping visual spaces.

Bright lights in Urbanplanen

Urbanplanen is a common housing association a few kilometres south east of Islands Brygge. It consists of prefabricated architecture constructed as part of the former mayor of Copenhagen, Urban Hansen's, drive in the 1960s to solve the rising population in Copenhagen. Today around 2,500 households and 5,400 people live there, but since its construction the area has increasingly been the centre of social problems. Part of Urbanplanen has been designated a 'ghetto', due to the fact that more than 50 per cent of the population originates or descends from non-Western countries; more than 40 per cent of residents between the ages of 18 to 64 being outside of the workforce or education system; as well as more than 2.7 per cent having been sentenced for violating weapon or narcotics laws.[2] In 2010, an extensive renovation project was initiated, including façade renovation, where light and colour played a central role. As Slater, Entwistle and Sloane have also shown in London,[3] social housing is often dealt with through engineered solutions of surveillance light and anti-social behaviour interventions, compared to more affluent areas where aesthetics is more prevalent. This move between fully lit social housing and darker affluent areas helps enforce and maintain ideas about what social housing is.

According to informants and media coverage, the problem in Urbanplanen is that some places *feel* unsafe when darkness descends, particularly during winter, and when walking alone. In order to make the area feel safer, the housing association has had some buildings painted with red façades and red light to create a more varied visual expression, and also as identification markers for each area. As a solution, a park area running through Urbanplanen has been renovated by cutting down some of the hedges and having more light installed as 'light is something most residents are missing [thus] more lamps are installed in the darkest places of the park'.[4]

Yet it is important to note that unlike many other places where light is lit as engineered solutions, the local residents' associations are promoting initiatives using light, not only to see, but as an activity that 'gathers' people, places and things. Local light festivals, such as *Lysfest* (Light party) and *1001nats lys* (1001 night lights), have been introduced in a collaboration between the municipality, police and local partnerships that aim to show that the green areas of the housing 'are as safe to be in during winter as in the brighter months, even if they may not feel that way to all'.[5] The point is not only to see how light may shape such feelings, but by making the light event an activity where people can meet to 'break the ice and bring residents closer together [. . .] it is about making the darkness safe and to create dialogue'.[6] This is done on several occasions; for instance: by lighting 1,001 candles along the footpaths; inviting lighting designer Hannah Anbert to light specific areas and create installations; having activities such as football matches lit by glow sticks; staging light-sword battles for the children; bread baking at a fireplace; DIY lamp making course; even a fire dancer; and getting free light-time switches from the police as burglar prevention.[7] Light is a central feature of making people feel safe. In this case it is not just as a visual technology, but as centre of a practice of getting to know one another and making the home look occupied. What is needed, beyond the current focus on light and individual psychology, is the understanding shaped above of how 'feeling safe' is also about being together with light, which then, in effect, makes people feel safe.

It is in such a context that we see how the inside and outside come together in ways different to that at Islands Brygge. When you walk through the neighbourhood of Urbanplanen, talk to people and see the different colours and light in the area, there is an awareness among my informants of how light makes it look 'multi-cultural'. As discussed in previous examples, there may be a sense of atmospheric community in such diversity, even if not everyone complies to a stereotypical cosy lighting. Yet in this area, such sense of community was not expressed, but needed to be installed. The renovation wanted to highlight that everything is not just uniform, although determining the meanings is difficult. This is exemplified by some informants saying that the red colour made it look like a brothel. More than seeing how public spaces are transformed, and lighting activities are used as a way of shaping a community and safety, the following shows how the diverse uses of domestic lighting come into being and make sense to the residents. Below, I highlight four households' different use of light to exemplify the diversity in how bright light relates to everyday practices and cultural logics, rather than how domestic practices relate to notions of *hygge*, sharing and community as in the previous chapter. It begins with Aisha, who personifies some of the sentiments shared by the other households as well.

The privacy of curtains

Aisha is in her early thirties, a mother of five, and was born in Denmark of Pakistani parentage. When she and her husband moved in to the apartment years ago, they hurried getting settled, and since she was nine months pregnant she just wanted to make the home child-friendly. Since everything was in a rush, they just needed some light and put a few strip-lights up in the living room. She talks about these strip-lights as 'neutral', even if they may seem bright compared to *hyggelys* in the previous chapters. It is 'what you put up if you are not so interested in light', as she explained. Lighting design was not on her mind. From interviews elsewhere, it is clear, however, that even the informants on Islands Brygge who did not care about, or invest much in, light, would never just set up a strip light. There is a sensuous judgement embedded in such choices.

Aisha's pronounced lack of interest in lighting technologies was not, however, reflected in her awareness of light. She had a strong sense of letting light facilitate privacy and actions in the home. In the living room two layers of curtain are installed, one made from laced and one made from thick non-transparent material. The laced curtains are *always* drawn. The curtains create a permeable border zone, which, according to Aisha, decorates and creates *hygge*, emphasizing the multiple settings people may describe as *hyggelig*. It also illustrates how visibility through the window is layered with different curtains, and that she felt uncomfortable with a clear, undisturbed view. For Aisha, privacy is central and light plays a key role in shaping such sense of privacy and vulnerability: 'when the light is on inside, people can really look in, and then I pull the heavy curtains.' Otherwise, people can see in, and the darkness outside makes the illuminated indoor space reflect in the windows, hindering her from having a clear look out. There is no reciprocity. In contrast to Chapter 5 (where pulled curtains on Islands Brygge were seen as a way of shaping an atmospheric community and bringing the exterior luminosity inside), Aisha felt it unpleasant that people she does not know can see what she is doing.

In this respect, Aisha confirms the findings from Kirsten Gram-Hanssen and Claus Bech-Danielsen (2012) in their work on homes among first- and second-generation immigrants in Denmark on the importance of drawing the curtains to obtain privacy. Aisha's curtains act as material medium for negotiating the embedding of oneself in social collectives (Garvey 2005; van der Horst & Messing 2006), where visibility becomes associated with vulnerability and various curtains therefore act to qualify privacy (see also UK example among elders in Ebbensgaard 2017: Ch. 7).

Making do

Aisha does not really like the strip lights, she just had not had time to change them, and actually, it turns out, it is not just the strip lights; in the kitchen hangs a chandelier. She is not satisfied with that either, and would like to replace it for one that gives a different, brighter light. At the moment, it is a golden one but Aisha wants one that is more 'Danish'; that is, she explains, a white one instead of gold. Gold is something more 'foreign', she finds. While the home is comparatively bright with the strip light, she sometimes thinks about having candlelight or a small lamp in the corner. She has candles but they are locked away because she fears they might cause her children injury. Aisha notes how, 'It would be *hyggeligt* in the evening sometimes, if you are sitting alone, to have a small lamp instead of the big lights on. That would be good. But for now, we take the kids into account for a few more years.' In this way lighting is entangled in ideas about danger, risk and what the future brings. In many of the homes visited elsewhere, even if there were small children about, candlelight was still a central feature of the decoration and by no means seen as a risk – it is something children are raised to be aware of. This is most potently illustrated at Christmas time, where tradition holds that families put up a spruce tree with Christmas decorations including tens of lit candles, and dance around it with the children before unwrapping the presents positioned beneath. But this is not the case with Aisha.

Aisha has energy-saving light bulbs in the kitchen, but she shares the frustration of my other informants about their poor quality: 'You can almost finish cooking before the lamp is all lit. I used to have them in the toilet and bathroom, but you could use the toilet and go out again before it was fully lit, so I stopped doing it.' While the incompatibility of lighting technology and practice – slow ignition and the act of going to the toilet – is clear to see, the 'inconvenient' bulb was still being used in the kitchen. Aisha noted that 'they take three to four minutes to ignite, so I turn them on, go back into the living room and watch television, and then go back to do what I wanted to. It is irritating.' In this way light becomes a way of structuring the use of space; it is about anticipating actions, but also about not acting upon the things that irritate.

Besides how light is used to structure space, what is interesting is the way Aisha recognizes and even accepts the poor quality of lighting in her home. On the one hand, she sees the strip lights as neutral, but she also longs for a lighting that is less neutral and more 'Danish' by tapping into the discourse around lampshades, candles, and white design – or at least not exclusively gold. For now, the bright lights make sense to her, but at the same time do not, as she actually finds visual reference points in the more subdued lighting found in homes such as those on Islands Brygge. It illustrates that while the home is centre for forging subjectivity, it is also clear that a home may not be exactly living up to one's ideals or wishes – in this case to adopt a more 'Danish' lighting, colour and *hyggelig* scheme.

Illuminating activities

Aisha comments on how the children know that it is evening when the parents light the strip lights and draw the curtains, and it is morning when they are pulled open. Pulling the curtains open is part of following the daily rhythm, of letting light in in the morning and shielding the privacy of the home in the evening. Turning on strip lights and drawing the curtains creates a different atmosphere, because the children 'know that they are allowed to run and jump when the curtains are pulled, or the light is not on'. When the light is on, they are quieter and can sit and watch television. When the lights in the street are turned on, the kids know that they should come home – at least during wintertime, because in the summer it sets around 10 p.m. A key point is that orchestrating illumination helps to shape the intensity of activities at home.

Another informant who uses light to illuminate certain activities is Sahar. She is in her late thirties, was born in Denmark of Pakistani parentage. As a housewife and mother of four, who spends a lot of time in the apartment where they have lived for three years, she is very aware of how she uses light to define the activity of the room. For instance, there would only be soft light if she is tired and needs to relax. But like Aisha, the household has mostly installed strip lights, which does not allow for that option or luxury. To create a soft light in the living room, she turns on the strip lights in the kitchen, which is separated from the living room by a half-wall. In this way, she 'borrows' light from the kitchen to use in the living room. In her bedroom, she has both a strip light and a small light, where the strip light is mostly for when she is doing housework. Sahar uses a lot of light, she says, switching on the strip light in the living room if the children have to do homework or they have guests, compared to when she is alone and is more relaxed. Sahar has a chandelier in the living room, but it is only used when they have visitors. It creates patterns on the wall, which she likes, but is not turned on when they are home alone. She explains how during weddings in Pakistan they light the couple's house with garlands. The chandelier is very expensive, and was offered as a gift and sign that people wish the couple a 'life in light', emphasizing the symbolic power of quantities of light.

When they have visitors, there are usually a lot of people, and 'that *of course* means that we need a lot of light'. The family turn on the chandelier, the strip lights, small lamps built into a glass cabinet, the strip light in the kitchen and a bowl lamp above the kitchen sink. She does this 'both to create the atmosphere and then, of course, you are not supposed to just sit and relax as I do normally. You have to be more awake. And light is what makes you more awake. The more the better.' Sahar is, in this sense, *qualifying* lighting through *quantifying* the social.

FIGURE 6.1 *Strip light home. Photo by Matilde Lykke.*

Candles are only used for birthdays, as a surprise for her husband or children, as they offer to her 'a good loving atmosphere' in the home, emphasizing the point of seeing 'lighting as care' as illustrated in Chapter 5. They can never be used when there are more guests, since people are talking too much for it to be a relaxing atmosphere. Having many visitors is not about *hygge*, but being alone is, to Sahar. What is interesting in this case is the way light is so entrenched in practices of being together. Her taken-for-granted notion that if more people are together you *of course* need a lot more light – although not candlelight – stands in sharp contrast to the examples from Islands Brygge from earlier on, where it was the *distribution* and not *quantity* of light that is the main point. Most of my other informants have been raised with an awareness towards the flame; it is not something to fear when visitors are around. However, unlike Aisha, there was no sense of Sahar's light being temporary; this was what lighting should be, and it made sense to her.

Making a new start

Zainab and Mohammed are a Danish-born couple both with Moroccan parentage, who moved in to the apartment in Urbanplanen a month and a half ago with two infants. They did not bring anything from their old apartment – they just wanted to start from scratch by buying new things. In this process, both Zainab and Mohammed showed a particular affectivity towards light. 'I really like light,' Zainab said. 'I get really depressed if it is dark, and feel helpless, ill. The more light, the happier I am. My life is best when there is really sharp light, it is the same outdoors – sun, sun, sun'. Not just any light will do, however. Zainab really detests some of the energy-saving light bulbs they have bought. Not just because they ignite slowly, but because she feels that she 'cannot find' herself emotionally. The family had not completely settled in yet, but even if they did put emphasis on the electrical bill, there still needed to be a lot of light. Yet the quality was still important and they despised 'hospital lighting' such as high kelvin light. Zainab had insisted that they got the right light for places such as the toilet, trying three or four different bulbs, and ended up replacing the lamp altogether, as it did not offer enough light. They illustrate a trial and error approach, with no sentimentality or nostalgia towards their possessions. They wanted to start anew, and in doing so, they could try different lamps and bulbs that are validated by sensing whether they 'can find themselves' emotionally in the bright light they emit.

They liked to *hygge*, and used candlelight to create a relaxed and romantic atmosphere at home, but they would normally have a lot of light, particularly if friends came over to eat. Later on, if they were looking to create a more intimate atmosphere, they would turn the lights down, so people's faces were not as discernible. Zainab and Mohammed had been brought up in Denmark by Moroccan parents. Although the light was bright compared to other homes described in other chapters, they did not share their parents' lighting culture with sharp lighting from few strip lights, unlike Sahar. In fact, they had noted the peculiarity of their friends' lighting practices when partaking in Pakistani weddings, or visiting multi-cultural homes with coloured lights and garlands: reds and yellows and greens, as if they were having a party, they noted. But Zainab and Mohammed wanted to start anew, making their own sensuous home.

The discussion with Zainab and Mohammed illustrates the role of the past in people's homes. Many people may not be satisfied with their home – such as Aisha – in part because they have to share it with their family, but also because they are burdened with things from the past that they have not had money, time or will to replace. Zainab and Mohammed had taken a radical step and simply not kept any furniture from their previous home. They now had the ability to start over and, in doing so, they could establish their own lighting culture, including the use of candles for *hygge*, which was not about adopting their parents' use of one strip light per room, but neither about fully emulating the subdued lighting seen in Islands Brygge. Their new home had shared elements of both: bright, but distributed.

A bright home is a clean home

Jasmine is a thirty-year-old student from the Philippines, who came to Denmark when she was seven. For Jasmine, a home is supposed to be organized and clean, and she is very aware of her use of light to shape what she called a 'clean home'. Some of her friends think that her home is too clinical and sterile, because all her furniture, carpets and bed linen are white. One of her South American friends even goes as far as stating that she does not feel at home in Jasmine's place, because 'she thinks it is too sterile. To me it is clean and comfortable. It cannot be too clean. That is perhaps why I have the white furniture because it symbolises some sort of "it is clean in my place".' In the Philippines, where her family is part of the upper middle class, she explains that light and lamps are used actively to show economic wealth; light is about prestige. She thinks people in Denmark also use a lot of light, but it is dimmed and distributed, and more often associated with homeliness than with prestige.

Although Jasmine feels it can be *too* dark with candlelight or *hyggelys*, she often lights candles, both when alone and when she has guests. She feels that it is a particular Danish thing. Although she does not think she lights that many, her fellow students with international background often tell her that the number of candles is quite excessive. 'They ask, "why do you have that many?" I just do, it is normal: Like buying bread, milk, and light.' She often engages in cross-cultural encounters and thus also has had discussions about the meaning of light. Some people visiting her have had the idea that with the many lit candles it is supposed to be romantic, but it is not at all like that, 'it is creating an expectation that it is *hyggeligt*, that "this is the way I *hygger* [verb – with the connotations of caring] around you"'. She uses the argument that there is so little natural lighting in Denmark that she needs to use white furniture and also puts mirrors opposite windows to exploit the little light entering the apartment: 'because we lack light, we have to create our own at home.'

Jasmine sometimes uses coloured lighting when she has guests; a slightly red bulb in some lamps, to offer a 'warmer atmosphere', adding colour to her white furniture. She also used to have a light blue glass lamp on the floor but that was only lit when there was an actual party in the apartment, since that colour is difficult to associate with a proper home. She is thus explicit about her lighting practices and preferences:

> Poor lighting is a single lamp in the middle of the room. Where light is right there but not in the rest of the room. It creates shadows, and you can only be active in that one part of the room. And then it is bluish, compared to a good light source, which has different kinds of light, distributed across the room, with different lamps.

She uses light quite deliberately, to set the scene or get things done. To her, *hyggelys* is great if you are reading and just need to relax, and the cold light from the energy-saving light bulbs are far from cosy. She says that it makes you think 'right, I need to get this and this done'. She cannot just relax in that light, 'it makes you more active'.

Here, bright light and large quantities of light are not only about number of fixtures but about the choices of material and colour of the furniture and walls. Jasmine's lighting practices are, in a sense, very much aligned with the central features of *hygge* lighting, even if quantitatively much brighter, which to her cross-cultural ensemble of friends is too excessive. It is a lighting practice oriented around shaping relaxed atmospheres, yet also very much aligned towards cleanliness. Cleanliness in Jasmine's home is not simply about getting rid of bacteria, but likewise an aesthetic cleanliness of whiteness (El-Khoury 2006; Schmidt & Kristensen 1986).

In the above four households, the reference to a stereotypical Danish light culture, including *hygge*, has been repeated, even if *hygge* is understood in quite different ways. What the examples share is a particular way of appreciating what, in other homes, would be seen as brightness – the bright white light that made Janni wonder if the inhabitant had been tortured at some point in their past. What the above examples have also illustrated is the diversity and variation in the details and structural aspect of why people have, and use, bright light. It is used to shape visibility, privacy, homeliness, order, cleanliness and certain activities. The examples show a tremendous heterogeneity. People are reflexive about how materials mediate their sociality, and live multiple and ambiguous relationships through light, where they both interpret and appropriate *hygge* and simultaneously being different to it in many ways, through personal history and memory. In the following, I present a final case, which is more oriented towards homogeneity, and how bright light comes to make sense. To do this, the chapter now moves away from Urbanplanen to the Middle East, among the Bedouin in Jordan where I initially became interested in light cultures (see Bille 2017, for a more elaborate analysis). The key point is to understand how lighting not only rests on diverse meanings, but also enables particular practices, sensations and values to change or be maintained.

Shine a light

Faisal's reception room is recognized as one of the most stylish ones by most people in his village in southern Jordan. Around 350 people live in the village, which was constructed to settle the otherwise semi-nomadic Bedouin tribes living in the area, of which Faisal originated. Faisal is a father of four grown-up children, a devout Muslim performing his daily prayers and a successful herdsman and land-owner. His home is often centre for guests. Hospitality is important in Bedouin culture and rarely has a family had no guests during a day, so having a good reception room is essential.

Muslim homes in the Middle East are built around the role of modesty, privacy, and hospitality, and this is no less so among the Bedouin (Othman et al. 2015). A house among the Bedouin generally has a reception room where guests sit, adjoined by a bathroom, kitchen and at least one extra room. In the reception room Faisal has spread some carpets on the tiled floor, a few mattresses and arm pillows, a frame with a quote from the Quran, a tea can and glasses for when guests are there, two strip lights and, from the ceiling, a bare, compact fluorescent light bulb, shedding bright light across the room (Figure 6.2). This is a beautiful room to the Bedouin, which, with the bright light and absence of furniture and objects, stands in sharp contrast to a Danish home, even the one in Urbanplanen. Faisal could afford to have it differently, if he wanted, but the bright light is not about economy, it is simply the right way to light a room.

In Jordan, the energy-saving light bulb has been welcomed with open arms, and there has been a rapid replacement of the incandescent light bulbs. When I asked Faisal and others why they preferred the energy-saving light bulb, with

FIGURE 6.2 *Example of reception room. Photo by the author.*

high *kelvin* level around 3,000, compared to the reddish tone one from the incandescent light bulb, which is around 2,700 *kelvin,* the common reply was that, 'it gives a clean light'. In Faisal's house, like elsewhere, the light is furthermore enhanced by the bottom 1.5 metres of the wall being painted with acrylic paint. The use of strong bare light bulbs, white upper walls, brilliant acrylic paint, and unfilled interior spaces exposes every corner of the room, and creates the impression for someone seated on the floor that the room is more spacious than it actually is. The visual illusion of an expanded space ties to the notion that the bigger the reception room, the more people the owner knows and therefore offers hospitality to. Among the Bedouin, prestige is gained through signs of generosity and the breadth of one's social network; the more people one knows and offers hospitality to, the higher one's social prestige (Meneley 1996). Light and the reflecting surfaces are not merely *reflecting* prestige, but part of a material and sensuous consolidation of this prestige of the house and family. Where the distributed and subdued lighting in Denmark was about shaping intimacy and avoiding tensions, in Jordan it is about formality, a need to offer hospitality, and, by extension, materializing one's social position.

Beyond enhancing visual perception and social identity, luminosity also ties into religious ideas. Among the Bedouin, shadows and dark areas are thought of as places where the spirit (*jinn*) dwells and misfortune lurks, further suggesting the importance of staging illumination and avoiding shadows in the house (cf. Al-Salameen & Falahat 2009). In the dusk, misfortune is also prevalent, and Faisal's family would, for instance, light incense both as a pleasant smell and as prophylactic strategy. These spiritual dangers are real to the Bedouin (even if also contested in religious debates (Bille 2013c, 2010)). It has been suggested that in a Muslim context the light settings in mosques are designed to emphasize the unity of the worshippers by fully illuminating all parts of the interior (Antonakaki 2007), and while not expressed by my informants, I see the analogy fitting to domestic uses of light and space. As Christian Borch notes, 'the feeling of unity might be moulded by means of careful sensory orchestration' (2015: 54) and this is no less so in the domestic setting. Lighting is part of establishing atmospheric communities, in whatever form and temporality it may take.

Light may be both a physical phenomenon and a spiritual experience, and it has been argued that the experience of light may serve as 'a bridge between the physical and spiritual planes of our experience [. . .] Physical events are spiritualized for us, and in visionary illumination the real nature of objects seen in the world disclosed' (Kapstein 2004: 1). Within Islam this complementarity of light and spirituality is reflected in the vast number of references to it. One of God's 99 names is 'Light'; the minaret of the Mosque literally means place of light or lighthouse, and a common metaphor is that God is light, while the Prophet Muhammad is the lamp (Campo 1991: 174, see also the Koran 24:35). Furthermore, more traditional ways of warding off evil spirits and deeds make use of light, mirrors, or the colour green, which are put on doors and windows alongside the name of God, which in itself is seen as prophylactic.

But it does not end there. And this is where cultures of light become central to understand. Over the last ten years, heavily green-tinted windows have been installed in most houses in the cities and villages in southern Jordan, including in Faisal's house. When the sun shines through the green windows, it bathes the walls, interior and people in a verdant haze. One is essentially confronted with the experience of lightness (Böhme 2017: Ch. 20). The green light exemplifies how defining an object is not merely a matter of delineating the tangible boundaries, but also how things are 'ecstatic' – step out of themselves and tincture their surroundings, as discussed in Chapter 4. When sunbeams interact with the window at the right intensity and angle, the window imposes its amalgamated nature onto other objects, and, in turn, completely alters their appearance. On sunny days, as the exterior extends into the interior, Faisal's reception room would be embraced by green light, and on cloudy days with less, but still noticeable. Only in the evening did the green light cease to occupy the rooms, leaving the window simply a green surface. Inhabitants would draw curtains to the other rooms, but few would do it in the reception room. The electrical light inside the houses meant that the streets in the village would be shaped by green light spreading from the windows of the homes. According to some informants, the spirits (*jinn*) would potentially be present in the village during the night, and light would scare them away. The very colour of green would also act as a 'reminder of God' as creator and protector. By installing God in body and mind, one becomes protected, which is a practice also seen with rosary beads and thanking God before eating, starting the car, and meeting people. The idea is that by attuning oneself to God one comforts emotional distress, establishing mental peace, and through God, safeguards spaces and people.

Faisal would not make any elaborate description of bright light as those from Urbanplanen did; bright light was simply taken for granted as the optimal 'clean' light. To interpret Faisal's use of light is to relate to cultural norms of generosity, honour, and domestic and spiritual safety. The bright light was entangled in manipulation of visual perception that would incorporate the meanings and practices taking place in the reception room. The sudden introduction of green-toned windows only emphasized the role of protecting and creating unity within the home, while also being cherished as a colour of spring when flowers and grass grow in the area. What has been striking is that while Urbanplanen displayed a tremendous heterogeneity, in how brightness is achieved the Bedouin homes were strikingly similar. But the point is that bright light is not just bright light; it makes sense in different ways.

Conclusion

From the five examples, we begin to sketch out broader issues about lighting practices, and the logics that underpin them. One of the interesting aspects for the purpose of this book is how the very act of sensing and shaping atmospheres is an 'atmospheric competence' that can be learned and developed (Albertsen 2012: 70; Böhme 2006b: 50–1; Grasseni 2004). In Jordan, the means of shaping the hospitable and safe atmosphere was relatively homogenous, while in Urbanplanen there were very different approaches to shining light. For some, shaping a different homely atmosphere is a desire, but one that has been temporarily suspended for the sake of children or surplus of mental resources. For others, starting anew offers this potential. It is about learning what the home is or should be. What is interesting in the cases of informants raised by immigrant parents, is their different emphasis on the role of *hygge*. For some, lighting the home in a cosy way in Denmark 'reflects a skill, talent, and concern for the values of conviviality and "positive sociality"' (Borish 1991: 278). But unlike most of my other informants, who are taught one such atmospheric competence in the name of *hygge* during their upbringing, it is different here. It is a bodily way of knowing and sensing the world in atmospheric ways that taps into ways of using the home. While 'the informal skills that facilitate subsequent competence in *hygge* [which] is an important part of the socialization process' (Borish 1991: 277; Levisen 2012: 100), it is also clear that many other things are at stake, such as vulnerability, a sense of cleanliness, the undone home, or bodily attunement to God for protection. The bright light makes sense, in this way, to the informants, either as 'neutral light' or as that which is appropriate to their friends and family and the atmospheric practices learned during their own upbringing.

The degrees and meanings of brightness are not only different ways to achieve the same values of community, sharing, collectivity; they also involve realizing entirely different values, be it cleanliness, prestige or religiosity at the centre of their 'light culture'. Lighting is thus a competence that can be learned, no matter if the goal is relaxation, activity, a sense of cleanliness, or visually materialising prestige and protection. To such ends, luminous brightness is the means. The previous chapter was very much about attuning oneself and the home through light, connecting space, creating a neighbourhood feel as part of an atmospheric community: vague, scalable, learned, but nonetheless felt. This chapter was more about diversity in meaning of similar lighting. Of how lighting preferences may be shared in practices, but not in meanings, but also how cultural premises entangle such meanings. The following chapter turns towards the knowledge people actually have about light, especially in a technological transition period.

7

Ignorance and illumination

Introduction

The transition to energy-saving light bulbs has forced people to change century-old knowledge about light. For some it is easy, as illustrated in the previous chapter: The light offered from the compact fluorescent bulbs available around 2012 makes sense. Others have been hard pressed to adopt a new visual comfort and knowledge regime. LED and compact fluorescent bulbs have forced us not only to sense differently, but also to *think* differently about what light is. With the incandescent light bulb generally available in supermarkets, the consumers knew what kind of light quality they would get from the highly standardized bulbs. To illustrate this sensuous knowledge, an informant noted, after I turned off the recorder, that 'you know, even in your body, what the light from a 60-watt bulb is'. It was a sensuous norm, one that neatly aligned with cosy atmospheres. In that sense, the focus on the bodily character of atmospheres is, as also discussed earlier, in need of nuances, where cognitive ways of knowing and not knowing play a larger role. Few would know that a 60-watt incandescent bulb actually meant: a 2700 *kelvin*, *ra* 99, c. 800 lumen, c. 98-watt heat emitting, 300° non-directional beam angle, dimmable bulb. Actually, terms like *ra*, *kelvin* and *lumen* may have been completely unknown to the average consumer ten years ago. These things were 'black-boxed' – made invisible by the term 60 watt. Aside from the shape of the bulb (normally the A-series, chandelier, or reflector bulb), and selecting a large or small Edison screw base, the choice was basically how much light you needed, defined then in terms of energy consumption, *watt*.

Things are changing, though. Consumers need to save energy – *watt* – and to do so have to stop thinking about light by proxy as *energy*, but more correctly reconceptualize what light *is* in terms of quantity, namely *lumen*. The energy-saving light bulbs have enforced a conceptual shift from what the technology *consumes* to what it *produces*. This marks the end of watt, at least as a way of understanding light.

It also entails a need to know the quality of light one is looking for, or rather the quantity of light, and for what specific lamp, as new bulbs come in many different shapes and sizes. Information campaigns and conversion tables in shops help show how many *lumen* a certain amount of *watt* equals (Figure 7.1), but the 'black box' of watt has been opened and the consumer now needs to know about other parameters: the *ra* (colour reproduction), *kelvin* (colour temperature), ignition time (as full illumination may take up to a minute), whether it is dimmable, and whether the size and beam angle of the bulb fits the lamp, since the energy-saving light bulbs are often larger than the incandescent light bulb. The size difference has led to critique for not fitting the Nordic lighting tradition of atmosphere and *hygge*, illustrated by the emphasis designer Poul Henningsen put on avoiding the glare from the bulb in his famous PH5 lamp, which later shaped the core of Danish lighting tradition. Throughout Denmark, particularly in public institutions, such aesthetic traditions may be sacrificed for the sake of energy consumption, but it often elicits judgemental comments.

Moreover, as part of a much larger socio-technical system, the new energy-efficient LED technology has also allowed for new sizes and shapes of lamp designs to be created. While heat emission from energy-saving light bulbs is markedly lower than an incandescent light bulb, some of the new lamps are constructed in materials that only work with bulbs, equalling, for instance, less than a 10-watt bulb, many of which would not work with dimmers (at the time of writing).

In this transition to new lighting technology, failed purchases are common, and such failure acts as a barrier to transition to energy-efficient lighting technologies, even if better technologies are now available. Many of my informants had often bought the wrong bulbs, had become dissatisfied with the light from their lamps, or simply continued using the incandescent light bulb (cf. Jensen 2013). Let me offer two examples: first, Martin, who laughed when he described his failed purchase of energy-saving light bulbs for the kitchen. He would often move to the kitchen, turn them on, leave again, counting '1. . . 2. . . 3. . .' and a minute later, return. Only then would it be fully illuminated. It was really annoying, he said, but he also felt too lazy to do anything about it, and it had been too expensive to just throw out; rather live with poor lighting then, or use his lack of knowledge, wasted money and wrong purchases as justification for not changing other bulbs. Failure leads to apathy.

Second, at the other end of the scale was Marianne, who a long time ago had bought quite a few energy-saving light bulbs. She had used them for so long, she said, that she does not even remember what the old incandescent light 'felt like' – and not just looked like. But she had to admit that the one she had in the ceiling gave what she called an uncomfortable 'shadowy' and 'heavy' light that left 'an oppressive atmosphere' rather than *hygge*, in the room. In the past, some visiting friends would even turn off one of her lamps, because they said the light it gave reminded them of a horror movie. But to Marianne it is too cumbersome to change everything again and she has got used to it anyway.

Both Martin and Marianne display an element of light apathy in response to the new technology, also because lighting is so embedded in everyday practices and thus difficult to change. The home needs to look and feel a certain way, and delayed response to actions – such as the one-minute ignition – or the 'cold light' from energy-saving light bulbs – is detrimental to getting the right atmosphere in the home and allowing certain practices to take place, such as reading, cooking, or visiting the toilet. Martin and Marianne's failures, and experience of lack of knowledge, raises questions about the role of knowledge gaps, and how people tackle such failures.

This chapter marks a deliberate shift in topics. It is no longer the atmosphere, *hygge* or meaning light is part of that is of main interest. Rather, the chapter is about knowledge, which is central to the *continuous practice of attuning atmospheres*. No practices of attunement without explicit, embodied, or tacit ways of knowing. I pursue two strands of analysis: one line of argument explores how lack of knowledge fosters emotional responses of frustration and even rage where material objects become the vehicle for expressing anger. A second strand of analysis goes beyond exploring what people know or not, and what it makes them feel and do, and explores how competence and incompetence becomes a performative strategy in social life. Here, I follow McLuhan et al.'s (2014) exposition of what Robert Edgerton has termed a 'cloak of competence' (1967). The cloak of competence is a social strategy used to display authority – even when it is missing – or manage stigma to appear normal, as shown by Edgerton among people with intellectual deficits. By reversion, McLuhan et al. argue that there is an equally social potency or motivation in appearing ignorant or upholding incompetence – a 'cloak of incompetence'. I use this concept to show how lack of competence and knowledge become part of the implementation and resistance to the energy-saving light bulb as an element in attuning atmospheres. The basic question asked is: how does lack of knowledge influence attunement of atmospheres through light?

To answer this, we need to discuss what ignorance is and how it comes to matter in social life. Here, light in the domestic realm is used as case to understand the affective relationship between human and technologies slightly shifting away from the atmospheric focus in the previous chapters, but also concentrating on broader perspectives of both the introduction of new ways of knowing light and the more general human interaction with technologies. But first, let us set the scene with an illustration of how the transition from incandescent to energy-saving light bulbs is not only a material and technological change, but also involves an atmospheric and conceptual change where ignorance and frustration of the visual effects lurks backstage.

Bring in the specialist!

The frustration people in Denmark have experienced when replacing the incandescent light bulbs at home was perfectly illustrated in the Danish TV program *Penge* (Money). Aired in February 2012, it contained a special feature on the new lighting technology. Representing the somewhat average consumer, we follow female pensioner Jette Rybak from Ølby on her quest to change all the bulbs in her house with the new energy-saving ones, either CFL or LED. The basic argument is that a lot of money and energy is potentially saved over time, even if the initial expenditure is high.

Visiting a supermarket and DIY shop, we follow Jette's frustration as she tries to find the kind of bulb she needs: 'Not as easy as buying milk and toilet paper,' she remarks. She takes great care in selecting one particular bulb to be used over the dinner table, and notes how the selected one is remarkably heavy, and then complains about the price tag of 200kr c. 27€. She bought 19 bulbs for around 1000kr, hoping that in time her expenditure will be returned in the form of savings. Returning home, we follow her as she inserts the new bulbs. The one in the kitchen slowly lights up, reaching its full potential after about a minute. She scoffs: 'absolutely horrible. I do not at all want that. It is simply too wimpish [*fimset*]. It gives a miserable light.' Clearly it did not offer the accustomed atmospheric light. She continues into the toilet, complaining about the light here as well: 'I don't think it is suited for a lady in her best age as me. I wouldn't be able to put on makeup, which is what I use the mirror for. [. . .] that bulb has to be returned.' It appears that in most cases Jette had simply bought the wrong bulbs, even if she was actually aware of the need to think differently about light as lumen instead of watt, when replacing them all. Most disappointing was the expensive bulb above the dinner table. It could not be dimmed, and she had specifically looked for one that could. Hence, even though she felt she was fairly competent in what she should look for with the new bulbs, she was greatly disappointed. The scene ends with her opening a closet where she had stashed a good portion of incandescent light bulbs.

Jette's bad experience illustrates the growing frustrations among Danish consumers immediately after the bulb ban. It is, at the time of writing, still a market with little standardization, quality fluctuations, and the odds of returning home with poorer lighting is high – although LED bulbs are rapidly improving. Instead, many people start with the halogen bulbs, which do not save as much energy as the CFL or LED bulbs. More than illustrating the lack of quality, however, the program also had the aim of informing the public of the proper ways of making a home-like or cosy atmosphere with the new technology.

To this purpose they brought in a specialist and project leader Simon Chræmmer from the Danish Energy Agency's initiative GOenergi (GOod energy) to discuss Jette's needs and select the appropriate light bulbs. A bulb is not just a bulb, so while one level of *ra* may be acceptable in the hallway, it will not work in the toilet. Jette apparently had bought 32 per cent less *lumen* than she had when she thought of light in terms of *watt*. Simon then showed that the packaging for the bulb above the dinner table did indeed state that it was not dimmable. Instead he had brought others, more suitable bulbs, with the proper *ra*, *kelvin* and *lumen*, for the specific place they were located, although for double the price. The main event at the end of the programme was the new dimmable bulb above the dinner table. Simon demonstrates that the bulb is indeed dimmable. It works! But alas, he is left with another problem: Dimming contacts come in many different shapes, and the one in Jette's house is of the sort that makes a loud hissing sound when used with that specific type of bulb, thus, Jette was back to the beginning. Slightly embarrassed, the specialist accepted that he did not know what kind of dimming she was using, and by the sound of the lighting technology, it became clear that the new one would not be acceptable.

This example illustrates that even if consumers think they know the different parameters of the new lighting technology, there may still be aspects of the technology they did not know about or take into consideration, such as 'the passage points': contacts and dimmers. Jette was clearly frustrated when confronted by her own ignorance; how could she miss the fact that the packaging said that it could *not* be dimmed? This blunder aside, she was still not impressed with the quality of light from the new technology. As she noted in an interview after the programme, 'I bought some terrible lighting. The bulbs took way too long to turn on, I had problems reading in the light, and it wasn't very cosy – almost like a Hitchcock-movie.'[1] After the specialist had brought better bulbs, she was actually satisfied. Yet there was a double layer of ignorance at play in the programme, showing how not only the consumers, but even the specialist, ended up illustrating the difficulties of the new regime of knowledge being implemented. The specialist from GOenergi exemplified that one should not assume anything about the lighting system, since without very detailed knowledge of the specific system, it was impossible to buy the right bulb.

One way of approaching the introduction of new lighting technology is to explore the practices and networks of actors establishing what energy-saving light is and is used for (see Jensen 2014). This highlights how the incandescent light bulb is seen as 'normal', and the energy-saving ones are 'alternative'. Their introduction also leads to new, *less* energy-efficient lighting practices:

people leave the lights on, install more lamps such as spot lights, or need more heating as a result of less heat emission from the new bulbs.

Another approach would be in terms of understanding different ways of knowing, reiterating the informant's sentiment above: 'you know, even in your body, what the light from a 60-watt bulb is.' Jette showed us that this sensuous habit and knowledge is not met with the new light bulbs. One needs to know light in a different way than previous notions of *watt*, and this has often led to failed purchases: wrong size bulb, less light, poor light, or simply malfunctions with dimmers. The consumer may blame the poor quality of the light bulb or lack of standardization, but often they may also only blame themselves for their ignorance: they should have looked for the sign that the bulb can be dimmed; that the ignition time was shorter, that the size did not fit, that the *ra* level was too low for a make-up mirror; that the bulb only produced the number of *lumen* compared to 50 watt and not 60-watt bulb; if the lamp says 'max 25 watt' equivalent of 250 lumen, can you use a 25-watt LED bulb equivalent to +1300 lumen? Such concerns can be seen as a knowledge gap; a gap that can be filled through public service announcements, information sheets in supermarkets, and television programmes such as *Penge*.

However, six years after the ban, there are still many failed purchases, despite media attention on the new conceptualization of light. According to lighting designers I have met, it is the lack of standardization of these bulbs that has led to a market flooded with poor quality bulbs. Many shops now display the lit bulbs, but it is often difficult to imagine what they would look like in one's home. Implementing a new technology involves a process of learning. Several recent studies on learning have shown how knowledge is shaped *through* technologies and practices (Hasse 2015; Sørensen 2009: Ch.4). Knowledge is not simply something in the mind, but emerges as situated knowledge, where experience, embodied engagement, and skilled vision is central to understanding the importance of concepts such as *ra* and *kelvin*. With the new light bulbs and the white-boxed light as *lumen*, *ra*, *kelvin*, etc., informants have become slightly disconnected from the lighting. People are used to experiencing light as withdrawn from explicit view; that is, something phenomenologically transparent that people do not see, but see *in*, or what Heidegger would call *ready-to-hand*. When it is different, light turns to a discrete visible object as *present-at-hand* (1996: §16,68).[2] The energy-saving light bulbs with low *ra* levels or high *kelvin* make the home seem 'sterile', the term used to describe the opposite of *hygge* by informants. The new light is no longer something 'you know in your body', and thus also creates a crack in the home as a *lumitopia* of integral light and place-identity.

As Mark Brosnan remarks, in terms of technophobia (fear of technology), 'until technology "becomes invisible" it will be found to create feelings of anxiety with certain individuals' (1998: 2). Similarly, post-phenomenologist

Don Ihde highlights this relationship between different kinds of knowing, when stating that 'any new technology in relation to human praxis, before it can become transparent and thus fully accommodated, must be "embodied" if it is to be "known" at all' (2010: 125). From these insights, there is a need to recognize that the new technology is establishing a new way of knowing light, which has installed a gap between what people may know cognitively in terms of *lumen*, *ra* and *kelvin*, and the embodied and atmospheric way of knowing light that the incandescent bulb and watt offered.

The primacy of ignorance

When researchers write about people's uses of technologies there is an inclination to uphold what I call a 'competence bias'; people either are, or about to become, knowledgeable and more competent users. For instance Nicolas Govoroff's (1993) inspiring study of hunters use of different size rifles, starting with small calibres. As the hunters' improve in skills, the calibre increases until the hunters become fully competent. Hereafter they gain prestige by their ability to down-scale again and kill larger animals while using less fire power than others. Competence in a traditional understanding becomes knowledge put into action, where knowledge is about knowing, skills is about ability, and competence is about doing. Lack of knowledge, skills or competence is a gap to be filled, since only thereafter can the user choose to use other technologies. The competence bias assumes either the knowledgeable and competent user, or if not competent, then how the user could *become* so by learning more or by adapting the technology to user competences through usability measures. Highlighting such competence bias is not meant to take anything away from the insights offered. The problem is, however, that the competence bias, even when it shows the social prestige of the competent hunters, glosses over the function and sociality of ignorance and incompetence (cf. Gregson et al. 2009).

There are many things people do not know, when it comes to technology. I have mentioned the term 'black boxing', which is a term often mentioned in the slipstream of explorations of the way technologies and knowledge are produced (Bijker et al. 1987). Bruno Latour describes it as

> the way scientific and technical work is made invisible by its own success. When a machine runs efficiently, when a matter of fact is settled, one need focus only on its inputs and outputs and not on its internal complexity. Thus, paradoxically, the more science and technology succeed, the more opaque and obscure they become.
>
> 1999: 304

People may recognize such black boxes in their everyday life, in that most often they have no clue about how something works, or how to fix it. Even ordinary things such as a car engine, radio or refrigerator are for many people far beyond their technical comprehension, and such items may only lose their mystifying power if they learn how it works (cf. Woodward 2007: 67–73 on Roland Barthes' work). The notion of opening the black box is used by Latour as a way for researchers to understand the processes through which technologies are constructed and come to matter. Latour's work on black-boxing is part of an *epistemological project*; understanding what scientific knowledge is, and how it is constructed. This helps us understand why, for instance, a technology ended up like it did, rather than in a number of other ways (cf. Cowan 1985; MacKenzie & Wajcman 1999; Shove & Southerton 2000). Inspiring as it may be, it still leaves *lack of knowledge* as something that can be filled and obtained for the sake of knowledge. To understand the episode with Jette, and the specialist, Simon, what is needed instead is to forefront lack of knowledge as an aspect of human–object relationships, and as a consequence the 'white boxing' of light by taking it from *watt* to *lumen, ra, kelvin*, etc. In that sense, the chapter continues the interest from Chapters 3 and 4 on vagueness and not knowing, even if there it was about vagueness of atmospheres and *hygge*, not something tangible as a technological object.

When James Ferrier coined the term 'epistemology' (1854: 46) as the study of the nature of knowledge, he not only defined a major branch of philosophy in counterpoint to 'ontology' – the study of the nature of being – he also highlighted a third philosophical term: 'agnoiology', the study of the nature of ignorance. Ontology and epistemology now have central place in all academic disciplines and educations, while explorations of agnoiology have only recently flourished (Croissant 2014; Firestein 2012; Gross & McGoey 2015; Rescher 2009). In its most basic form one can see ignorance as the absence of knowledge. But there is a distinction in many languages between absence and lack of knowledge. Lack of knowledge implies a privation of, say, something that should ordinarily or naturally be there, whereas absence implies that the knowledge is not there. But this may further rely on different motives such as when people prefer not to know, or when there is a complete ignorance (Haas & Vogt 2015 also term 'presumed' and 'investigative ignorance'), illustrating how there are the known knowns, the known unknowns, and the unknown unknowns (High et al. 2012: 16).

Such philosophical clarifications are needed, yet few have examined the social aspects of ignorance (Chua 2009; Dilley 2010; Gershon & Raj 2000; High et al. 2012; Hobart 1993), and the way material objects mediate such ignorance has received even less attention (cf. Bille 2013c; Højer 2009; Pfeil 2012; cf. Norman 1988). These studies clearly show that there is more to ignorance than simply a blank slate or vessel ready to be filled. Ignorance forges identities, is part of political struggle, and is actively produced. Proctor and Schiebinger (2008) highlight the political implications of the deliberate, cultural production of ignorance. They deal with the way ignorance and doubt are induced; for example, by the publication of scientific data that aims to cast doubt on, nullify or suppress knowledge claims – most famously the risk of cancer from tobacco use. A key point for Proctor and Schiebinger's approach is that ignorance is always a product of cultural and political struggle. From this position, studying human and technological interaction is about investigating how this relationship is politicized through networks of casting doubt, denial and secrecy. Clearly, it is not just a top-down process of those in power, but also about resistance from ordinary consumers, who may disagree or choose to trust other sources as evidence. The point is to see how a process of casting doubt, or using ignorance, is a political act.

Piers Vitebsky (1993: 101) argued that knowledge and ignorance are less *cognitive* than *evaluative* terms. They are used to assess other people, and are centrally bound up in denigrating alternative ways of knowing. This insight is taken up by High et al. (2012) who explore the social uses of ignorance, and the different logics, ethics, and emotions it incorporates. They note how it is easy to suppose that the people under study have the same 'desire for knowledge, and the same aversion to ignorance [. . .] with the result that situations in which ignorance is viewed neutrally – or even positively – have been misunderstood and overlooked' (2012: 1). There is a danger of mistaking ignorance for stupidity, error or confusion. Their point is to avoid seeing ignorance as a *lack* or *absence* of knowledge but rather allowing it to be a topic with its own logics, motivations and effects.

Yet another perspective is offered by Roy Dilley (2010). He shows how it is possible etymologically to make a distinction between the related terms 'ignorance' and 'nescience'. Ignorance relates to the suffix *–noscere*, meaning to 'perceive or apprehend' by the senses, compared to *nescience*, related to the suffix *–scio*, meaning to understand: knowing by the mind. This is exactly what Ihde is also referring to, with the notion that technologies are only truly known when they are embodied. As with Jette, my informants may often know what terms such as *lumen* mean, but that knowledge has not been embodied. Even with the actual insertion of the bulb in the lamps at home, their cognitive knowledge is put to the test by their embodied knowledge. The lumen from a 60-watt bulb is 'truly known', while the equivalent c. 800 for an LED bulb, are purely numbers. Similarly, the 60-watt also automatically meant 99 *ra*, while the new technologies actually have a number for the colour reproduction, depending on each model or brand of bulb, and the effect of this number is only realized when it is tested.

But technologies not only signal ignorance, they also help to create it. Proctor and Schiebinger, High et al., and Dilley, have offered vital starting points for understanding the social aspects of ignorance. I wish to pursue another line of enquiry by focusing on how people may find themselves not only ignorant *about* a certain topic, but also how this evaluation and positioning of ignorance is constructed *through* using and conceptualizing material forms, and the social positions this ignorance helps to forge. The role of technologies in establishing and evaluating ignorance becomes particularly interesting when new technologies are introduced and there may be moral attributes to knowing in the right way. For instance, by knowing that the CFL bulbs are to be recycled due to the mercury content, rather than simply thrown in the bin. Such moral aspects appear in what may seem as an excessive waste of energy; they may be seen in the current emphasis on recycling, up-cycling, or second-hand shops, where the ethos is an obligation to think about the potential value of what one throws away. But it is also apparent in questions about *how* one consumes, and here lighting may offer another illustrative example. We will therefore now turn to one of the main seasons for artificial lighting, namely Christmas, and explore the way new lighting technologies have been rapidly accepted to illuminate the dark evenings at the expense of energy use. But Christmas illumination, while atmospheric and joyful, also takes skills and knowledge – and not all may have that as the later part will explore.

Illuminating Christmas lighting

On a quiet, dead-end street in the suburbs of Copenhagen there is a certain anxiety in the air. It is the Friday before first Advent Sunday of Christmas. This is the day where around 15 houses are lit to various degrees with thousands upon thousands of Christmas garlands, child-size elves, LED reindeers, and decorations can be found on roofs, flag poles, hedges, and garages. The lights automatically turn on every morning and evening and spread 'a Christmas atmosphere' onto the streets. At the weekends Christmas music even pours into the street from outdoor loudspeakers in a cacophony of secular, religious and popular themes. On this day, at six o'clock, a firework fills the sky and a few hundred neighbours and visitors begin to move from house to house to watch each household turn on their Christmas lights. As all the houses are lit, the congregations meet, eat pork sandwiches, drink beer and enjoy the lights and festive atmosphere. There is even free – with the option to pay, if you wish – *Gløgg* and *Danish Doughnut*.

Excessive Christmas lighting also happens in many other places in Denmark, either in individual houses or as seen here, as a group of neighbours who have started lighting their houses with excessive lighting over the last decade. As they have attracted more and more neighbours, tourists and press

FIGURE 7.1 *The joy of Christmas lighting. Photo by author.*

to the middle-class cornucopia of illumination, both visitors and house-owners make reference to Chevy Chase in the film National Lampoon's Christmas Vacation. Visitors and social media oscillate between astonishment and indignation. On the street, visitors comment to each other in various ways, spanning from it being a waste of energy, to being wonderful and joyful, to the kitsch Americanization of Christmas, etc. I was told by one house-owner that in some cases it has even led to vandalism, where installations are stolen, or, in one case, a brick was thrown through a window on a neighbouring street with similar lighting.

It is the one of the few areas where the LED technology has gained rapid success. Despite being blamed for energy waste, the home-owners claim that it is not that expensive. When I met up with one of the house-owners, one of the first things he did was to show me a detailed table of energy consumption in December, stating that with LED bulbs it may normally be around 650 kroner extra a month. The biggest expenditure is buying the new garlands and installations. For others, it is a more costly affair, at around 1200 kr., along with the cost of renting a crane some years ago to decorate a 14-metre high tree with 4,000 bulbs – 'but it's *hyggeligt*' is the argument. The tradition started with just one house back in 2005, and took speed the last decade.

As Tim Edensor and Steve Millington (2009) have shown, the extensive use of Christmas lights in the UK is about bringing joy and happiness, even at the expense of energy or middle-class taste. Lighting up houses for Christmas with multicolour lamps creates a particular geography of illumination as it predominantly occurs in working-class areas. In the media and internet coverage of the phenomenon, the excessive lighting in the UK is frowned upon, highlighting a particular working-class emblem of being 'chav' – a pejorative stereotype for lower-class, young people wearing 'bling'. The places where the Christmas lighting is showcased also becomes entangled in issues about the geography of class and social deprivation – you can understand the meaning and identity of a place by evaluating the Christmas lights is the idea. However, Edensor and Millington show that among the practitioners, the lighting display is more about conviviality, generosity and collective festivity. As one of their informants notes, 'To me it says celebrate Christmas. It's a gift to share. That to me is what Christmas is about, to appreciate what we've got' (2009: 113). It is a technology, and knowledge about using it, that helps shape an atmosphere, a sense of community and joy of sharing, is central to the argument in e.g. Chapter 5.

The same Christmas sentiments are apparent in Denmark, although the phenomenon is not exclusively associated with the working class. Rather, when it is a collective display between neighbouring households, it often appears in the middle class suburban areas of detached housing. Aspects of competition most definitely appear, most often with a degree of distanced humour at the expense of 'good taste', but people also make the decorations personal, fearing it may be labelled 'silvanisation' (*silvanisering*, after the DIY shop Silvan), if they just buy it all there. Having visited the street the last seven years at Christmas time, the diversity or support for the lighting has by no means diminished. One year a family had bought a house on the street during the summer, and were surprised when they saw the lights go up in December, and then the large numbers of people visiting the otherwise quiet street. Perhaps not what they expected, but a year later, they had already succumbed to the tradition, and two years later among the houses with most decorative lighting. They got caught up in the atmosphere, they said, which brings you closer to your neighbour and offers something to gather around both in terms of finding the right technology, but also in terms of creatively contributing to the Christmas light atmosphere.

Central to the Christmas illumination practices is the way lighting is talked about in terms of bringing and sharing joy. Stories abound of tourist buses arriving on the small street with Chinese tourists taking photographs. The fact that the cul-de-sac has a kindergarten at one end only adds to this, when you see excited, happy children wanting to stay and watch the lights rather than go home. In other places in Denmark, this emphasis on joy and altruism is highlighted by a donation box outside a house to support a charity.[3] It is for the greater good, and essentially (they claim), in the spirit of Christmas, despite what may otherwise look like a lack of taste and unnecessary environmental pollution. The lighting is, as Chapter 5 also showed, about creating atmospheres and a sense of community in the otherwise somewhat anonymous suburbia, and to materialize the Christmas messages of care, joy and benevolence. But it was also specifically about knowledge.

In the weeks prior to the festive evening in 2012, I participated with two families in installing all the cables, setting up the garlands, and making sure it all worked. The Christmas lights were stored in the loft or in a section of a shed, and the whole *séance* started with inspecting the different objects and installations about to be revived. Power cables may have been broken during last year's dismantling and storage, new items added, and new habits of positioning the lights may emerge, as each item is re-evaluated every year. The owners position the life-size snowman on the lawn, run the cable over the lawn and plug it in. The garlands have to hang a bit lower than expected because otherwise the cable cannot reach, and nails are fixed in the wall to hang the elves upon. As the amount of light has increased, a few practical measures have also been taken, such as installing extra outdoor plugs, extension cables, and extra fuses. Even this most mundane of actions – plugging in a power cable – requires a tremendous amount of know-how and experience. Too many lights in one socket may turn out to be a problem, but so can too few, as some automatic timers cannot detect small amounts of watts. It is a situated knowledge that implies learning that takes form through social networks and materials that strike back; they break, they do not do as the user thought, and through such resistance shape learning process (Sørensen 2009: Ch.4). After the above illustration of the implementation of LED technology, I want to return to ignorance and focus not so much on the amount of knowledge or material networks it takes to set up 50,000 light bulbs, but rather on what happens when either the technologies, or human knowledge, fail.

The emotional responses to technological ignorance

Back to the day when all the lights are about to be turned on, we meet Peter and Inge who are next in line – as people gather outside their house, preparing to clap, gasp and comment on the spectacular show that every house so far has shown. Peter runs in and turns on the lights, but people have hardly started clapping before one of the fuses goes out. Peter hurries back inside to replace the fuse; it gives a small glimpse of light, and then goes out again. He then runs around trying to figure out the problem; again, no success. Which garland is it that causes the problem? Meanwhile, standing in the street with many of the neighbours and visitors, the wife tries to explain her own passivity by highlighting how she does not understand electricity – it has always been Peter's job – and she is more into cooking and cleaning. The neighbours soon start making sarcastic comments about Peter's lack of technological skills, bringing back memories of his earlier encounters with defiant technology when buying garlands. It was not uncommon for my informants to have bought the wrong garlands, or put them in a place that irritated the neighbour and hence had to move them. Nor is it unusual to have these kinds of problems on this night. The sarcastic comments were all meant in good spirit, and brought laughter onto the streets.

However, the joyous atmosphere was not reflected in Peter's actions; he was clearly irritated by the malfunctioning technology, and his lack of competence to solve the problem in front of an audience. In the end, Peter found the damaged garland. Irritated, he pulled it out of the socket with one hard and aggressive, almost violent, movement, and inserted a new fuse. Finally, everyone rejoiced as the light came on again. But the technology had got under Peter's skin, emotionally, and he still did not know why it had malfunctioned; he only knew the solution.

The technorant

To understand the two brief but exemplary cases above – of Jette buying new bulbs, and Peter getting emotionally upset, trying to fix the lighting – we need to go beyond looking at it through the lens of *lack*; that is, as something missing. Rather, I want to view the technological ignorance displayed as a starting point for analysis in its own right, rather than illustrating a gap to be filled. There may of course be many valid arguments to suggest that knowledge of electrical systems, or better understanding of light bulb terminology, is preferred, but that is secondary to the argument here. The white boxing of light has made the different terminology of parameters of light quality accessible, yet having to know all these parameters also sets the scene for frustration, dissociation and new ways for social positioning. First, we need to understand the common experience of being technologically ignorant, or what I combine in the neologism: *technorant*. To most people it is a familiar experience of human–object interaction, particularly with the increasing number of modern technologies where the user is detached from the production phases and often approaches objects as black boxed. It is impossible to know everything about all the technologies we surround ourselves with. Some people may intuitively know more, of course, or be willing to learn more, but no one is omniscient. Technorant is not the description of a person, but of a relation established through acts of technorancy. We are all technorants; there is always something that has been black-boxed, or an object we do not know how to use, even by intuition. It may also be the opposite, that is when something like light has been white boxed and we may not fully understand all the new terms, or have not embodied the sensuous effect of the terms. It is a momentary condition, rather than essence of being, like being ignorant is not something one always is, and it may be a feigned ignorance, used as strategic device. If viewed through the lens of the technorant, I see (at least) five ways of human technology relation arising in failed interaction: *Technophobia*, *Exploration*, *Prosthetics*, *Animism*, and the *Cloak of Incompetence*.

Technophobia is the fear of technology. Unlike the techno-luddite, who does not want more technologies, the technophobic relation is one where fear and anxiety erupts with the use of technologies. Mark Brosnan has shown how such technophobia – in relation to computers – particularly applies to women, but is shaped in cultural ways of conceptualizing gender and technologies, particularly when comparing the West to the East. Technophobia comes in different degrees, from being uncomfortable, to seeming cool but cognitively negative on the inside, to being visibly anxious (1998: 13). It is not about knowledge or its absence, but about the emotions and trust shaped in the process of interaction and evaluation (1998: Ch.6).

Another version is what I term *Exploration*. It is somewhat the opposite strategy of technophobia as it involves engagement with the object. It is a scene where intuition and confidence reigns. To draw on Claude Lévi-Strauss (1966), it may entail two different ways of engaging the world: the engineer and the *bricoleur*. The engineer is the one planning the moves, while the *bricoleur* takes a more improvised approach, making use of whatever is at hand. This explorative mode is what we saw with Peter (at least to begin with), when he was still trying to solve a problem he did not know the roots of. The explorative mode does not equal success, because even though all efforts are put into making the technology do what it is supposed to do, one may still fail.

However, Peter soon developed a different relation with the light garlands, one I call *Prosthetics*. Peter allowed his lack of knowledge – and unsuccessful exploration of both engineering and bricolage – to provoke him emotionally. It was the emotional frustration that erupts when something is not working, and most often it is when you need it the most. But what happens when either technology or human knowledge fails? Frustration and disappointment comes with not knowing, and blame (including self-blame) is often extended to the decision-makers or repairman. Some grab and smash the object in anger; others keep tackling the problem in good humour; some give up. Psychoanalyst Serge Tisseron (1999) describes the emotional relationship between individual and object as the way in which the social significance of people and things are sustained. Tisseron argues that the emotions invested in the object, or that they trigger, are released through human practices with the object. More than simply an extension of the human, objects may have prosthetic qualities that are balanced by the inclusion within emotional states of those whose bodies, both sensuously and cognitively, they extend (Dant 2005: 62). We may hit the pillow in anger, or like Peter aggressively pull the cords in frustration. Peter tried to fix the problem with exploration and intuition – exchanging the fuse – but failed, and was ridiculed for his lack of knowledge.

A variation of this kind of relationship is found in what I call *Animism*. Alfred Gell has explored this type of relationship with his notion of 'vehicular animism' (1998: 19), that refers to how people attribute a soul to an object. Gell discusses the relationship people have with cars, and how they may be named and personalized. When this animistic version of the car is not working, it is seen as an intentional act on the car's part, at least momentarily. It is a fine balance and temporal relation between understanding the relationship as an animistic or prosthetic one, essentially defined by the attribution of intentionality and personality to the technology. This is where we see computer rage and other types of anger expressed.

The term technorant obviously combines the term technology and ignorant, but by putting emphasis on the techno*rant*, we may also emphasize the affective aspects of using technologies, particularly apparent in the prosthetic and animistic approach – the willingness or temptation to do harm to it by going on a *rant*, both verbally and in gestures. Practices such as fixing the light involve gestures and actions that 'not only enable us to accomplish actions, they accomplish them in a certain way' (Thibaud 2011: 209), where 'human gestures are movements of the human body with a meaning' (Albertsen 2012: 71). The way people gesture is infused with meanings, such as undertones of identity, emotions, or power demonstrations. Gesture thus 'concerns not so much the nature of the activity (the "what" of the action in the process of being accomplished) as its manner of execution (the "how" of the action, or the form it takes while being accomplished)' (Thibaud 2011: 209). How people and things interact is also evaluated in moral regimes of how one *should* interact with them. Kicking, ranting or violently pulling things is a sign of lack of control, and may be humoured, or displaying one's social position. The incident described above was not only a question of whether Peter got it to work, but also if it was done properly. More than merely ignorance about how to fix the light, copy machine, computer, or television, when the user's intuition is no longer helping, the prosthetic and animistic perspectives explore the more emotional aspect of the technorant.

Finally, there's the possibility that people are actually not ignorant, but simply look like they are. This has been called the *Cloak of incompetence*, which may relinquish 'the possibility of being seen as competent – with all the rights and privileges that go with that attribution' (McLuhan et al. 2014: 381).

The cloak of incompetence largely defines the strategic claims of ignorance, as we saw Inge avoid the embarrassment of not figuring out what was wrong, or when people justify not buying energy-saving light bulbs because they do not know what to look for. Unlike the above aspects, the cloak of incompetence assumes that there are some levels of competence that are just hidden. It is a way to dissociate oneself from providing a solution, but, as with Inge, the effect is also that it becomes part of solidifying gender roles. Consequently, ignorance is not only something used evaluatively to characterize other people, but also as a stance that may be deliberately adopted, since by claiming a position of ignorance, doubt or uncertainty, one may avoid over-committing oneself to any particular social current. People may thus 'play dumb', deliberately avoid or fail at doing a task to help reach strategic goals (Kotamraju 2011; McLuhan et al. 2014: 371). Mistakes are also often excused if people did not know any better in comparison to cases where they should have known. We see this social role of incompetence when elders use lack of knowledge in installing a television and the like, as occasions for having visits from friends or family. McLuhan et al. notes that such 'displays of incompetence or competence are best viewed, not as dichotomous processes, but as the ends of a presentational continuum' (2014: 363).

Conclusion

The point of this chapter has been to show how a new regime of knowledge results from the introduction of the energy-saving light bulb within the domestic realm. Unlike the other chapters that have shown how practices of attuning atmospheres takes place, this chapter has focused more on knowledge and ignorance about a technology as part of attempts to attune a space or oneself. With Jette it was the impossibility for her to have the same light and thus atmosphere as before, only more energy-efficient lighting, and with Peter and Inge it was about the problem of contributing to the luminous Christmas atmosphere. Rather than looking at how such knowledge is implemented or old knowledge made obsolete, I wanted to highlight how it is not always fruitful to focus on the competent, skilful practitioner, or the way he or she becomes one. Or more precisely, the skills it takes to recognize and orchestrate *hygge*, illustrated previously, was confronted with the lacking skills in knowing the technology – cognitively and bodily. A focus on the role of ignorance also allows us to grasp how the socio-material life of orchestrating atmospheres is structured around ways of not knowing; the gender positions it creates, the very notions of knowledge it shapes, and the emotions it stirs. It is about the skills it takes to attune spaces through lighting technologies; it is about understanding that the Christmas light show establishes a sense of community through an atmosphere of joy and celebrations – and a bit of excessive kitsch; And it is about understanding the way bodily and technical knowledge is challenged when engaging with lighting technologies. Don Ihde's (2010: 125) notion of a gap between an 'embodied' and 'disembodied' knowledge, highlights how the new lighting technology has not helped overcome the new regimes of sensation and terminology. We should look for the ways ignorance and the link between *not-knowing-by-the-senses* and *not-knowing-by-the-mind*, in certain instances, may become a dominant theme over knowledge in that it allows people to position themselves and evaluate others in the continuous practices of shaping the way they see and sense the world.

In this way, this book has now moved from the socially embedded nature of light (Chapter 2), over the culturally informed atmospheres it helps shape (Chapters 3 and 4), to the routinized practices of shaping atmospheric sense of self and community (Chapter 5), and the diverse premises and logics of bright lighting of homes (Chapter 6). This chapter has shown that when attuning atmospheres, technological ignorance and incompetence are not simply a cognitive lack of knowledge, but also a means of evaluating things, other people, and affirming one's self-identity through the use of material objects. The following chapter broadens the scope further by addressing the politics of lighting technologies and atmospheres.

8

Reflections of sensory politics

Introduction

There has been a close metaphorical relationship between morality, spirituality, truth, and light since earliest historical records (Blumenberg 1993; Brox 2010; Kapstein 2004). The illuminated light bulb is a symbol of 'a bright idea'; in literature and film, darkness represents a mental and emotional abyss; and religious vocabulary abounds with references to light. In the twentieth century in particular, there has, however, been an intensification of moral judgement regarding the polluting energy consumption spent on light on a more concrete level, which, in turn, has reversed the metaphorical association between purity and light.

Prior to the International Climate Convention in Copenhagen in 2009, Thor Pedersen, a prominent member of the ruling party, Venstre, and former Finance Minister, proclaimed that the climate debate, and initiatives to reduce CO_2 emission, was getting out of hand and interfered with people's personal freedom, adding that he enjoyed returning to his empty house and seeing the lights on, coming from his windows. The remark about intentionally lighting the house when not at home, and therefore wasting energy, was met with moral indignation in the press. It was used by political opponents to question the government's intentions and abilities to promote Denmark as being at the forefront of environmental sustainability, contained in the Danish promotional slogan 'State of Green'. Yet the remarks also resonated with the increasing scepticism in Denmark about the moral duty of reducing CO_2 emissions. It was not so much about the scepticism surrounding the 'facts' of global warming, but rather the way in which the minutiae of everyday life were evaluated through CO_2 emissions. And since light holds such a central role in shaping domestic atmospheres, as attested to throughout this book, small variabilities make a huge difference. Thor Pedersen's comment was about atmosphere in every sense of the word: it was about the earth's atmosphere, because the energy consumption used for lighting makes up a large percentage of global CO_2 emissions; it was about a political atmosphere, where opponents were searching for ways to blame the government at a time where its political skills were on international display; and it was also about a more experiential atmosphere, of what the home *ought* to look and feel like for it to be a proper, welcoming home. Homely atmospheres trumped environmental atmospheres for many Danes at the time.

The past three decades have witnessed the emergence of ethical consumption as a large-scale economic potential (Carrier & Luetchford 2012; Hobson 2006; Lyon & Moberg 2010). When it comes to fair trade or organic food, the ethical aspects of consumption are very much up front: paying a fair amount to workers in the developing world, or making sure animals have more space and fewer chemicals before we slaughter and consume them. People may choose *not* to buy these products; this may be a decision based on price, or on habits, and not necessarily on the qualities of the organic, fair-trade or energy-efficient goods. Yet when people do not adopt a new technology, banning the old one is a powerful way to change habits. This then in turn may lead to rapid improvements of the new technology. It is a political step where the majority, or those in power, impose their preferences or political goals on others. In the case of the energy-saving light bulb, a critical member of the European parliament from Germany went so far as to call it 'light bulb socialism'[1] – banning a technology rather than letting market forces push the companies to develop more efficient and attractive bulbs.

Steven Chu, US Energy Secretary and former Nobel Prize winner in physics, recognized this occasional human deficiency or inability to change behaviour, and the need therefore to take political steps. In 2011 in the context of out-phasing the energy-saving light bulb, Chu stated that: 'We are taking away a choice that continues to let people waste their own money.'[2] This kind of rhetoric and political action shows that a lighting technology is by no means neutral or value-free; rather the transition in lighting technology is closely associated with ideas about personal freedom, morality, care and climate change.

In Denmark and Sweden in 2007 and 2008, this moral position on energy-wasting technologies was taken a step further by *Asfaltjunglens Indianere* (Indians of the Asphalt Jungle). As an activist movement, they had on several occasions let down the tyres of four-wheel-drive cars. In an act of protest resembling vehicular animism (Gell 1998: 18–19), they were in a sense constructing the SUV as a monster not worth caring for (cf. Bellacasa 2011: 90). Upon entering the era of energy awareness, wasting money through inefficient technologies is, or should be, a moral abomination to some of the critics, and political or activist steps must be taken to eliminate such waste. Such politics were also enforced in the recent normalization of the non-smoking sensorium with the smoking ban in public spaces across the globe, rapidly making people aware anew of the sensory environment, leading them to reflect and perhaps even be disgusted by earlier norms. Earlier habits are seen in a new light where 'pollution' emerges as a category through which one's actions are classified.

Following David Howes and Constance Classen, this indicates that 'perception is shaped by culture and that society regulates how and what we sense [which] is also to say that there is a politics of the senses' (2014: 5). What counts as dirty, clean or polluting, is thus not only shaped by personal preferences but also by the social values it carries (2014: 1). As argued by Peter-Paul Verbeek (2011), the ban of the incandescent light bulbs not only *reflects* but *materializes* morality through new technologies; that is, the technologies *themselves*, and not only the human action of lighting (or driving a four-wheel drive), are entangled in this moral position by materializing values. In the case of lighting, the politics of sensing the world has always been a central capacity (Dillon 2002; Nye 1990; Schivelbusch 1988). In so far as morality comes into a new light through technologies, it raises issues of the politics of lighting and also the need to frame a social approach to light and lighting technologies. This book has already shown the role of the embodied knowledge and practices – the collective meanings and cultural contexts integrated in the use of light – and this final chapter homes in on the politics of light as it becomes integrated in human lives, not least in terms of dismissal or rejection. The aim of this chapter is to discuss how technologies are entangled in sensory politics and materialize morality, and further to exemplify how sensuous experiences become normalized.

First, I reflect upon the historical process of normalizing the sensible; how technologies materialize morality, and how the quality of energy-saving light is disputed and the possibilities it offers. Following from this I explore the cultural notions underpinning the resistance towards such moralizing technologies in terms of the productive prospects of dirt and pollution. The final section then summarizes the overall objective of the book: to initiate a more social approach to lighting, one that integrates human practices, meanings, knowledge and politics to complement the technical, psychological and physiological approaches that have otherwise dominated research on light and lighting. Such an approach shifts the focus towards a better understanding of the ontology of elements – such as atmospheres and lived experience – in contrast to one that reduces living with light to disparate fields of individual objects, bodies or senses.

The emergence of a sense of hygiene

The field of history of ideas offers an interesting point of departure from which to understand sensory politics, with the establishment of *a sense of hygiene* in the global urban population from the end of the eighteenth century and particularly during the nineteenth century (El-Khoury 2006; Foucault 1973; Schmidt & Kristensen 1986). In Denmark, according to Schmidt and Kristiansen (1986), this sense of hygiene was more than just a bacteriological order in the home; it was also an *aesthetic* and moral order of how the home and world at large *ought* to appear, not least in the sensuous terms of smell and vision. As a consequential effect of the rationality of bacteriological avoidance, there was an increasing division of space into places of sleep, work, living, dying, and eating to assist such order.

But the sense of hygiene is not just about division of space; it was equally about installing procedures and codified behavioural norms and manners to introduce 'cleanliness' as an ideal in Denmark. Central to this is what Schmidt and Kristiansen call 'project housewife' (see also Olesen & Thorndahl 2004). The construction of 'the housewife' was achieved by the state, consumer organizations, and commercial agents collaborating in the promotion of a new sensorium – *the sense of hygiene* – based upon a scientific paradigm. It was also about creating a symbolic space for such a sensorium as the working class increasingly occupied urban spaces – what Schmidt and Kristensen argue marked the emergence of the modern home. It was about establishing a more formalized knowledge about our sense of hygiene and the home through schooling programmes such as the 'housewife schools' (*Husmoderskole*) in the late nineteenth and early twentieth centuries (1986: 116). The procedure to shape this sense of hygiene promoted four central features – sun, light, fresh air and cleanliness – and yet the correlative effect of this was to view housekeeping not only as a moral or bacteriological project, but as a *feminine* project (1986: 120–1). It was not just about *physical* uncleanliness anymore; it was about *symbolic* uncleanliness. The effect of this moral regime is still evident in Denmark today when one invites guests into one's home: from washing the toilet, or vacuuming, to the more mundane making of beds – despite knowing that those places may not be dirty, used, or represent a health risk. It was about:

> establishing a conception of the home as a secluded social space, also as an intimate social milieu, a symbolic space that entailed more than any other space – hence the importance of the housewife's *sense of domestic hygge* (cosiness).
>
> 1986: 125, my translation, original italics

The housewife thereby became a creator with a flair for the aesthetics of intimacy, which in Denmark came in the guise of the emic concept of *hygge*, and the central role and aesthetics of light, both natural and artificial.

However, the rise of functionalism in architecture in the early- to mid-twentieth century forced itself upon the housewife. The aim for functionalism was a rational and practical use of space, inspired by the functional kitchen shaped for, and by, the housewife – a use which should also be reflected in the living room. Functionalism offered an appreciation of the standardized and sterile. Nonetheless, this standardization threatened to eradicate the housewife: 'formalized and reduced to its functions, the housewife becomes replaceable and thereby herself a standard without personality' (1986: 159, my translation). Hence, it did not take long before domestic *hygge* once again became the task of the housewife 'incarnating individuality and personality in the midst of systematic uniformity' (1986: 159, my translation). It is also in such a context of home-making that we shall also see the rise of fashion magazines, lighting design and ideas about good taste emerge (cf. Olesen & Thorndahl 2004; Stidsen et al. 2014).

What is even more interesting is the way Schmidt and Kristensen map how an amalgamation of different regimes of knowing takes place, when the moral, aesthetic and health factors become inseparable in lived experience, whereby politics are embodied in everyday practices, perception and knowledge (a similar example is seen in Buchli 1999: 52, with Lenin's lamp in the Soviet Union). Understanding lighting in this way entails attending to the entanglement of pollution, moral, and visual preferences, where technology is not an innocent bystander.

Sensory politics

Melvin Kranzberg famously argued that technology is neither inherently good nor bad, but nor is it neutral (1986: 547). Technologies help, even force people, to act in morally acceptable ways. Speed bumps, seat belt alarms, and ultrasound scanners, are examples of technologies that have been seen as 'bearers of morality' in that they have consequences if we ignore them, or make us ask new moral questions (Latour 1992; Verbeek 2011: 2,12). Approaching technology in this way moves morality away from being solely a human affair into the realm of technology.

It is important not to disconnect sensory experience and atmospheres from the broader societal issues in terms of politics and power; what we may term sensory politics : How politics have consequences on how the world is sensed, and how ways of sensing become political topics (see also Borch 2014: 13; Edensor 2017; Fennell 2011; Schivelbusch 1987, 1988). Lighting is part of sensory politics, and so are atmospheres in so far as current political steps aim to change sensorial norms. It becomes political when the incandescent light bulb – which has otherwise illuminated most of the twentieth century as a potent sign of purity, welfare, progress and indeed moral character – is phased out by the governments. It is also sensory politics when consumer resistance in Denmark results in hoarding incandescent light bulbs. It is political when urban spaces are staged as theatrical narratives, pushing anti-social behaviour aside. It is political when social housing areas are swathed in light to engender a safer atmosphere. Illumination technologies are political in so far as they present an opportunity to frame the atmospheres that people live through; politics is therefore about shaping new distributions of the sensible creating norms, communities and exclusions (Rancière 2004), expanding the scope of neoliberal influence (Fennell 2011). Charles Hirschkind has in this broader sense of sensory politics, for instance, shown how a 'banal' technology such as audio recordings of Islamic sermons in Egypt have contributed

> to shaping the contemporary moral and political landscape of the Middle East [which] lies not simply in its capacity to disseminate ideas or instil religious ideologies but in its effect on the human sensorium, on the affects, sensibilities, and perceptual habits of its vast audience. The soundscape produced through the circulation of this medium animates and sustains the substrate of sensory knowledges and embodied aptitudes undergirding a broad revival movement within contemporary Islam.
>
> 2006: 2

As shown above, the sense of hygiene entailed making a *physical* problem of bacteria into a *moral* problem that changed the *aesthetic norms* – that is, changing people's values, perceptual norms, and ways of thinking. Much like the nineteenth-century instalment of a sense of hygiene, or sense of religious ideology, I would argue that what we see in the reconfiguration of thinking of light as lumen instead of watt, in Chu's moral claim about wasting money, and in commercials and the push for consumers to switch to energy-saving LED bulbs, is an initial attempt to establish a heightened *sense of energy* among citizens. In such a step, human practices and technologies are increasingly measured and talked about in terms of energy consumption and, by extension, its impact on climate. It is the continuous efforts made to remind us of the environmental effect of lighting practices. Take Earth Day, for instance, where light is turned off for an hour; this is a reminder of the amount of energy that is currently invested in illuminating our world, and that we once lived with much less light and need of light. There has undoubtedly always been an awareness of the cost of fuel – the amount of work put into chopping wood for a fire; the price of oil, gas or electricity, etc. – but with the rising awareness of the global impact of human consumption practices something new has occurred: an awareness that there are effects of energy consumption beyond the proximate environment, that our energy resources are not infinite and the impact on the earth is severe.

Some people are, however, reacting viscerally and affectively as this sense of energy enters the aesthetic domain. *The sense of energy* – as the amalgamation of energy-efficient technology, morality, and aesthetics – is, on

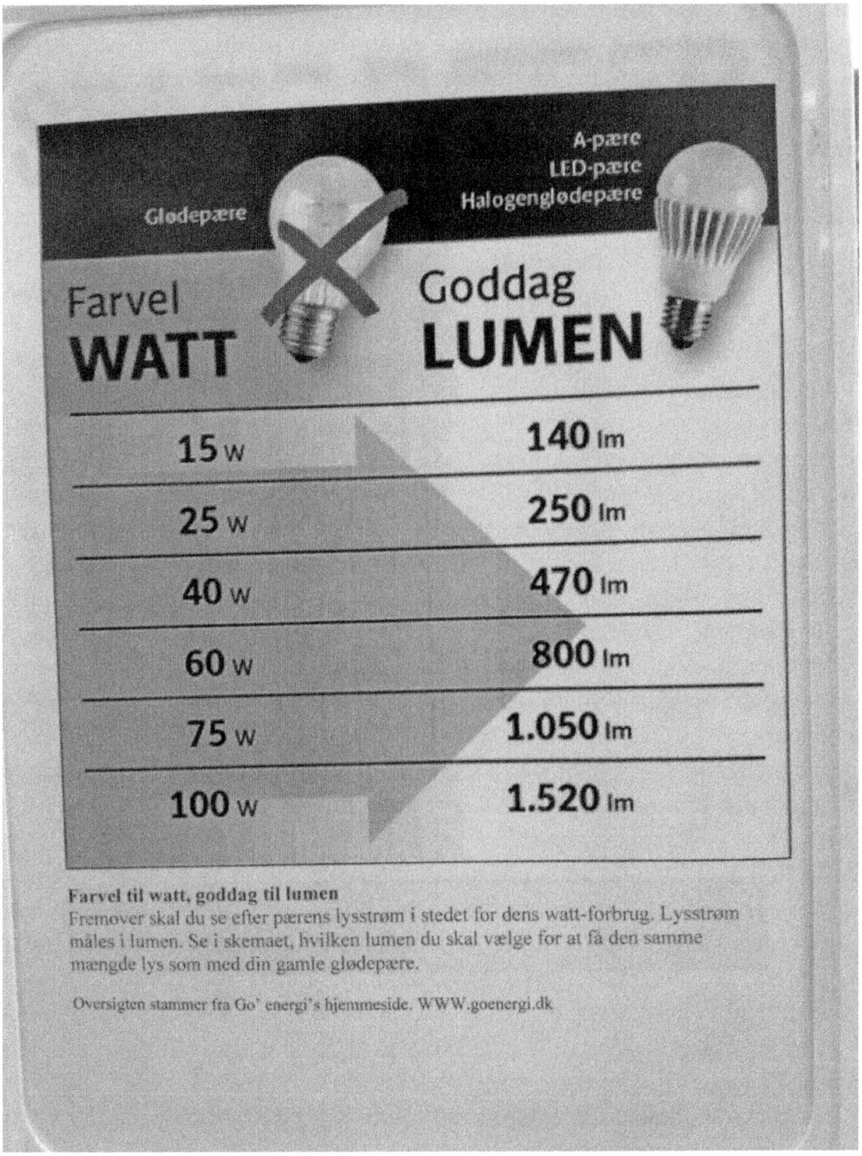

FIGURE 8.1 *Goodbye Watt, Hello Lumen. Photo by the author.*

the one hand, the empirical process of installing an energy awareness at societal level; it is about the political steps taken to lower energy consumption, such as supporting renewable energy, and making such awareness an integral part of organizational behaviour at societal level. On the other hand, it is also about engaging the individual consumer as part of the solution, politically promoting local goods, energy-efficient cars and technology, and raising awareness of the CO_2 footprint of people's practices. The underlying premise is the need to change both *behaviour* and the way consumers *think*, as discussed in Chapter 7, in terms of thinking about what is produced – *lumen, kelvin, ra* – instead of what it consumes: watt. It is about changing behaviour, but it is also about making sense of energy efficiency, both economically and environmentally. It is about articulating a moral obligation to care about saving energy, and materializing morality in new energy-efficient technologies, which also, literally, makes us *think* differently about light. The new lighting technology is not the end of 'watt-thinking', of course. However, watt is mainly used in terms of paying the bill, and not as a measure of light. What becomes the standardized reference is light itself, not the energy it consumes, which has led us to focus on what lighting *does* rather than what it *is*, what it *consumes*, or its climate *impact*. In so far as people adapt the new lighting technologies, they are no longer really tied to wattage; people orchestrate *lumen*, and in doing so they put themselves in touch with their environment. Lighting technologies have acquired the capacity to organize the public – visually, affectively, cognitively, and economically. But not without resistance.

In Denmark, many people have not been satisfied as represented in critique in the media and the effort to narrate guidance by government agencies, particularly in the early years around the light bulb ban. The aesthetic order has not changed. Simply translating watt to lumen does not solve the problem of experienced quality of light. What has proven necessary is rather a technology that adapts to cultural regimes of sensing. The energy-saving light has challenged our attunement to space, which has shifted the coordinates for reflection and debate from energy consumption to atmosphere. The new lighting regime, unstandardized as it was in the years around the 2012 ban, not only installs a different degree of sensuousness but an essentially different aesthetics of sensing. Consequently, aesthetics in relation to the sense of energy is not about surfaces or superficiality – in comparison to the ethics of environmental obligations and care – but rather it establishes a new sensorium which becomes the bearer of such ethics.

So far, the point has been that energy-saving light bulbs are part of materializing the politics of energy consumption. In the process of changing people's way of thinking, and their perception of what light is normal, the light bulb has also come to represent climate consciousness, by doing the energy-saving work. In a sense, the Danish politician Thor Pedersen could still light his empty house at night, but would be using more energy-efficient technology. He would not have internalized the sense of energy as a moral order, simply sustaining the dominant aesthetic order at a more energy-efficient price.

Elizabeth Shove has shown how ideas about comfort and convenience are changing through the interplay between politics, market forces and people's everyday lives (Hand & Shove 2004; Horta et al. 2014; Shove 2003; Shove et al. 2007; Shove & Southerton 2000). Instead of developing ever more technologies to fix rising energy consumption, she and her colleagues have called upon the need to view social practices as fundamental to curbing energy consumption, rather than resting on economic and psychological theories of behaviour where lifestyle and tastes are expressions of personal choices (Shove 2010; Shove et al. 2012: 2). The latter approach, Shove argues, sustains, albeit more efficiently, existing standards and conventions. Rather, she is proposing 'social theories of practices which emphasize endogenous and emergent dynamics' as a better response than the 'theories of social behaviour focus on causal factors and external drivers' (2010: 1279). What we need to understand is how human practices are nudged or changing. In extension, I claim here that the atmospheric elements of such practices are pivotal to our understanding, as they impinge upon, and guide, sensory politics or its rejections. In the same way as the sense of hygiene was not simply a bacteriological order, there is also the need to change an embodied level by normalizing a new visual aesthetics, and this has proven more difficult. While the moral undertones are often clear in terms of how to consume and the large economic incentives in doing so, the sense of energy differs from the sense of hygiene in the case of Denmark, when it comes to normalizing the aesthetics of the energy-saving light bulb – the Compact Fluorescent Lightbulb, in particular.

In everyday life in Denmark, the many examples from this book have shown how the new light bulb, particularly in the years immediately after the ban, did not succeed in 'becoming invisible' or 'embodied'. Jette (in Chapter 7) was clearly experiencing a gap between light and body, and Zainab (in Chapter 6) 'could not find herself' emotionally in the new light. Rather, you could say that in the home we have seen a shift from atmospheric perception where light participated by shaping lightness, into a way of perceiving where light stands out, detached and present-at-hand, rather than being integral to the homely *lumitopia*. In a sense, people start taking notice of the light when it is different, and – at least momentarily – construct it as an external technology, rather than integral, 'withdrawn' quality of space. From this position, one can perhaps suggest that in many cases the best lighting design is unnoticeable *as light*, because it captures the atmosphere that people are accustomed to, expect, or anticipate, appearing simply as a normalized lightness. That is, in the case of domestic lighting the lighting does not stand out *as* light, but as atmosphere. But this lumitopia where light is 'withdrawn' from awareness is however confronted by a technology that has not been embodied, and is thus still part of the perceptual field waiting to recede into a background.

While people in many places across the globe have adopted the new visual aesthetics – because it makes economic, moral and bodily sense or in terms of previous lighting traditions – this has not been the dominant discourse in Denmark. This is exemplified earlier when Nørretranders and Eliasson noted that: '[y]ou remember with your body how *hyggeligt* it is when you dim or turn off the evening light and light two candles on the table' (2015: 82, my translation), or the informant who knew 'even in your body, what the light from a 60-watt bulb is'. Rather, it has been a matter of suspending one's visual norms by either hoarding incandescent light bulbs, buying less efficient halogen bulbs that more resemble 'the old light', or simply making do until better bulbs are made. It is important to note that while I talk at a more general level here I do not intend to make claims about 'what Danes are like', as I have also shown the diversity in lighting practices and orchestration. The descriptions throughout the book nonetheless illustrate that the new light has installed a different *kind* of sensorium at a general level; an awareness of light, its flaws, and one's own presentation in it. This new sensorium is then adopted in different *degrees* depending on biographical, economical, and geographical conditions. Practices in the countryside, such as Thor Pedersen's, may not resemble those of the inner-city Islands Brygge, or among single residents, or people on low or high income. While this is a matter of *degree*, the general awareness of a new *kind* of sensorium is predominantly shared.

I met people who had fully switched to the energy-saving light bulbs (that is, excluding halogen bulbs), where they may have found satisfactory bulbs in IKEA, or bought well-tested or expensive bulbs in specialist stores. But it had rarely been without mistakes on the way, and *hyggelys* was the common key reference. With the exception of the residents in Urbanplanen with multicultural background, even those informants in Islands Brygge who stated that lighting did not matter and simply needed to be 'functional', would still not install fluorescent tubes or have an unshaded lamp. The visual comfort provided by shaded light was still present to some degree in the 'functional' light.

Only very few of my informants put energy consumption, rather than *hyggelys*, as their main priority. Charlotte was one such exception. Living a single life with very little financial surplus in her life, she was very explicit that saving money on light was a prime objective. Until recently, she had not been satisfied with her lighting at all. Some of her bulbs ignited very slowly with what she termed 'sleepy light'. Her daughter had found some cheap ones in the local supermarket, and Charlotte had reasoned that she would also have that luxury. She had heard that lighting had impact on the environment, although she mainly just wanted to save money. For instance, her kitchen is located next to the external walkway, from where public light enters her apartment. At first, she was irritated by this light, but then she began to like the fact that she did not have to spend money on lighting her kitchen herself. She reasoned that prioritizing economy over *hygge* lighting was something she had learned during her upbringing, and during her showcasing tour of the apartment she pointed out all the areas in the apartment where she does *not* have to turn on a light. For instance, she always left the doors inside her apartment open, thus borrowing light from other rooms. This includes the bathroom: 'what do I need to see, sitting there for five minutes alone? I wonder how other people can't see that they are spending money, and it is not necessary to have all that light.' Energy, climate and economics were closely related. But the relationship also became part of social validation and positioning, including moral judgements about how other people were wasting their money. The general awareness of a new kind of sensuousness thus becomes qualified at the level of social categories such as class, gender, marital status, age, ethnicity, urban or rural spaces.

Closing the debate

When the light bulb ban came into effect in 2012, the only technology that was sufficiently developed and available to common households was the compact fluorescent light bulb (CFL), if we exclude halogen bulbs, which are only marginally more efficient than incandescent bulbs. Many academics had already criticized it for its poor qualities, even to the point where some scholars argued that it had an effect on people's sleep patterns (e.g. Czeisler 2013). In 2008, the Danish Electricity Savings Trust (*Elsparefonden*) released television commercials targeting the 'myth' that the energy-saving light bulbs gave a 'cold' light, which has been one of the main complaints from Danish consumers reluctant to replace their incandescent light bulbs. More than spreading information about what people may not know (for instance that due to their mercury content the energy-saving light bulbs needed to be recycled), campaigns like this aimed more directly at battling competing knowledge claims. However, the CFL bulb can also be seen as an intermediary technology, since hopes are, and were, high for the LED bulbs as the future means of artificial lighting. LED bulbs now offer close to similar light, in terms of kelvin, as incandescent light bulbs, but back then in 2012 they were few and far between, and it is only recently that more people have started buying LED bulbs. In a representative survey by the *Energistyrelsen* (the government agency monitoring energy in Denmark) among Danes over 18 years old, they note that while only 13 per cent had bought an LED bulb in 2013, this had increased to 74 per cent in 2015, although the numbers do not tell us how many they had bought or where they used them.[3] Despite such increase, Danes still hoard the incandescent light bulbs and seek ways to buy them. This is perhaps best illustrated in another study done by

Energistyrelsen. They estimate that despite the incandescent bulb being banned in 2012, five per cent of bulbs bought in 2014 were still incandescent because shops and consumers used legal loopholes to import the bulbs, and the closest alternative, the halogen bulb – which is slightly more energy efficient than the incandescent bulb – is still on sale until 2020, when it will be phased out.[4]

At the time of writing, it is indisputable that most of the energy-saving light bulbs you buy in the nearest hardware store or supermarket offer a quantifiably different spectrum and quality of light than an incandescent light bulb – one needs to spend time meticulously sorting out the good ones from the bad in order to get a bulb that comes close. In the promotion of the energy-saving light bulbs – both compact fluorescent and LED – there has been an almost standard process of what scholars exploring the social construction of technology call 'closure mechanisms' (Bijker et al. 1987; Pinch 2009; Pinch & Bijker 1984). What we see here is a technological controversy, where there has yet to be a 'crucial' scientific, and not just commercial, experiment, showing that there is no difference (Pinch & Bijker 1984: 425). In lack of such 'scientific closure' on the most common bulbs in the supermarket, there has been a move towards other rhetorical closure mechanisms where the point is not so much solving the problem of no difference, but whether the consumers see the problem as being solved. Throughout this book we have seen examples of people dissatisfied with the quality of light, but who have not pulled themselves together to do anything about it. The 'problem' shifted from being accessing *hyggelys* to one of *economy*. It is a rhetorical closure where the problem is redefined: the problem is not quality of light, but the energy consumed. Here the aim is to convince us that this technological controversy is really not a controversy at all, because the problem is energy consumption, not lighting quality. Another example is how some of the older energy-saving light bulbs had *ra* level (colour reproduction) around 80, where the difference is markedly visible and people see it as a problem, but with *ra* 90 the difference may not matter as much, although in other contexts, for instance artistic or surgical light, it may indeed matter. The problem is solved, even if not solved. Yet it has been a slow process of convincing the Danish consumers since people's sensuous experience does not resonate the 'no-difference' claim. In Denmark light plays such a central role in home-making practices and ways of seeing, where nuances matter, that may not matter in the UK or Spain.

In essence, these are times where a debate has not been closed. LED seems to be the answer, but the new possibilities it offers in terms of colours, positions, and automation are by no means settled as the quality is neither the same as the incandescent, nor has it been fully normalized as the new normal – at least in Denmark.

Automated morality

From various strands of material culture and technology studies we learn that technologies do more than simply *mediate* morality; they actually *materialize* morality in order to do 'ethics by other means', showing people *how* to live (Verbeek 2011: 19). New lighting technologies are materializing an environmental morality; they are actively doing, rather than passively mediating, the moral labour of humans by taking over human environmental conscience as 'a material participation' (Marres 2012). But by doing so people run the risk of attending to climate *by proxy*: yet another area where we can free our cognitive capacity and leave it up to technologies to sort it out. This shifts attention to morality – as socially approved norms – to also being about the way technologies help constitute morality and care (Bellacasa 2012, 2017).

One such example of climate technology performing moral labour is the so-called 'intelligent lighting'. This may be part of large, software-based systems, allowing for remote control of everything from heating to lighting, or more simple room sensors, switching off light when no active movement is sensed, thus letting technology do the memory work. These lights are now present in many people's homes, particularly as an outside convenience and security deterrent, to alert the homeowner if there is movement outside, and also in places one rarely enters, such as attics or cellars. A newer addition, more or less exclusively in the public or work–life domain, can be seen in the window blinds that go up and down depending on how much daylight enters the room. Here scientific standards of 'good lighting' are orchestrating the luminosity in the room regardless of the practices occuring. The current trend is that not only the blinds, but the electrical light itself, dims or increases according to external light conditions. More recently this also includes the circadian rhythm light, particularly installed in areas people work nightshifts, that alters its colour spectrum depending on the time of day – the bluer end of the spectrum in the morning (to produce the hormone serotonin that awakens the body), and more red light in the evening (to produce melatonin that makes the body relax). In this sense, people are becoming less actively engaged in shaping their environment, and technology is lighting the world by retaining a specific amount of light according to standards of 'good lighting'. While acting as a responsive relationship between technology and human, automated lighting conceals the human origins by distributing agency to the technology, while humans have actually had a central part in constructing the categories and schemes underlying it.

This way of letting technologies do the labour has also been called the 'device paradigm' by Albert Borgmann (1984), where new technologies are implemented to cope with the effects of human thought, rather than changing the human–environment understanding. Instead, Borgmann proposes to distinguish between a device and a thing in terms of how we relate to it: A musical instrument commands skill, but a stereo does not. A *thing* promotes practices, a *device* promotes consumption. A thing, as he notes, is inseparable from its context, in that 'the experience of a thing is always a bodily and social engagement with the thing's world' (1984: 41). He exemplifies this with a wood-burning stove: 'A stove used to furnish more than mere warmth. It was a *focus*, a hearth, a place that gathered the work and leisure of a family and gave the house a center' (1984: 41–2). On the contrary, a *device* is a commodity that does not demand our skill, and becomes concealed unless it is not working. For instance, 'a device, such as a central heating plant procures mere warmth and disburdens us of all other elements' (1984: 42).

Automated lighting is like central heating: a device in the background of human attention, unless, of course, it is not working according to plan, such as when you have been sitting too still in a room with automatic lighting and it switches itself off. As has been repeatedly shown in this book, lighting is part of a practice of attuning the atmosphere; that is, the skill of attuning atmospheres to *hygge*, work, domestic chores, solemn reflection, party, etc. Drawing on the etymology of focus as *hearth*, he argues that practices offer to be 'the orienting force of simple things [that] will come to the fore only as the rule of technology is raised from its anonymity, is disclosed as the orthodoxy that heretofore has been taken for granted and allowed to remain invisible' (1984: 199). Borgmann talks about focal things that are inconspicuous; they flourish at the margins of public attention, but they are highlighted when they do not work, and are different. In the case of intelligent lighting, it is a technology that, while serving to fulfil the paradigm of well-being and sustainability, still materializes morality by implementing a device paradigm, rather than highlighting morality as a practice of material engagement and participation. In this way, we see a split between the informants' emphasis in words and actions on lighting as a continuous practice of attuning spaces and the lighting technologies' acquired capacity to automate and standardize lighting.

According to Borgmann, this distinction highlights a moral significance, where things constitute a commanding reality, in contrast to devices that procure disposable reality (1992: 296). It may of course thus also be a burden to keep the heating or lights on and not a pleasure. Bringing in Borgmann is not intended to offer 'living light' as a nostalgic contrast to electrical light; rather, the point is to bring attention to how people insist on investing time, money and skills in orchestrating atmospheres through lighting, even when they could have it differently. As Schmidt and Kristensen also point to, with their history of ideas approach to rationality, it may be more 'fruitful to pursue [a history of rationality] in the *social practices*, since they can be determined as social forms of ordering' (1986: 8, my translation, original italics). Lighting thus becomes a practice of attuning the world, and in doing so, making a *thing* out of what may be a *device*, by socializing lighting.

Dirt and light

We may call the lightbulb ban another political step in the rise of sensory politics (Cf. Fennell 2011), but one that was confronted by sensuous habits. What we see in Denmark is an appropriation rather than full implementation of a sensory politics. Thus, more fuller implementation of sensory politics has been seen with the smoking bans enforced across the world, rapidly altering olfactory experience and de-normalizing the experience of smoke (Bell et al. 2010; Dennis 2016; Tan 2016). The very idea of smoking in an office or airplane today seems distanced in most of Europe. But it is also a visual change in that the smoke was able to capture light, offering a different layer of depth and distance in a room compared to the tangible surfaces (cf. Schmitz 2014: 13–29). The recognition of the politics of new ways of perceiving, and the subsequent skills and knowledge this new materiality employs, also points towards the role of practices in understanding the uses and roles of light in human life. It is precisely the ontologically vague nature of atmosphere that make them such interesting phenomena, as they are unlocalizable, culturally embedded sensations that despite their vagueness are 'real' to people, and new technologies or laws may highlight such realness. Among my informants and in presentations in media, light has gone from being at the margins of attention and taken for granted in people's everyday life, to stepping forth as a technological device, and being a potential source of pollution.

In the same way as the question of what is to be considered the proper visual orchestration of space, addressed in the other chapters, so too is what is considered dirty, embedded in cultural understandings of normality. The social history of hygiene over the last 150 years has shown that the use of a guilt-inducing rationale for increased hygiene has required a reconfiguration of the senses, where dirt is to be considered simultaneously a moral and a physical issue (Campkin et al., 2007, p. 2; Schivelbusch, 1988; Schmidt and Kristiansen, 1986). Yet like atmospheres, so too does dirt slip 'easily between concept, matter, experience and metaphor' (Campkin et al., 2007, p. 1). To my informants, for instance, such categories as 'dirty' and 'clean' are not necessarily polar opposites, or a distinction between good/clean and bad/dirty. To them there is a fine balance between cleanliness and what is called 'sterile'. Some informants tell stories of how they clean the house before guests arrive; this is not necessarily because it is dirty, but just to feel that the room is clean. In Chapter 6 we heard how Jasmine keenly orchestrated her home according to ideals of cleanliness; it was not too sterile, but rather clean and comfortable to her. As she mentioned: 'It cannot be too clean. That is perhaps why I have the white furniture because it symbolizes some sort of "it is clean in my place".' If the moral duty to have a clean home is not achieved, then others, such as my informant Anne, describe how dimmed light becomes a tool, in that it, 'does not reveal details. Things are allowed to stand a bit in the shadow. Borders dissolve. It may sound strange. But I think, that if you put fluorescent light in here, I would be able to see all the dust bunnies on the floor. Those things disappear when you have the cosy-light.'

Light, in other words, is part of producing an image of a clean home. However, there is a fine line between the clean and the sterile, even in Scandinavian minimalism. The sterile, both as aesthetic expression and bacteriological reality, does not fit in easily with the orchestration of domestic atmospheres. If the home is *too* clean, *too* neat, *too* new, *too* planned, *too* designed, it would resemble a public office or hospital – not a home. In that respect, I would argue that the opposite of *hygge* or cosiness – at least in a material perspective – is not necessarily 'the un-cosy' or 'uncanny', but rather 'the sterile'. The sterile is a particular kind of sensuous encounter that both deals with regimes of knowledge about bacterial absence, as well as the very sensuous encounter with aesthetic and bacterial absences. The sterile allows for surfaces to step forward *as* surfaces containing the absent, and this is easily interpreted as a lack of personal touch on interior design and, by extension, lack of personality.

If dirt is 'matter out of place', as Mary Douglas (2002: 44) famously argued, the problem of pollution as a by-product of lighting traditions becomes more ambiguous. What in one regime of knowledge is understood as dirty – such as particle pollution from fireplaces or candles – is, in another regime, a product of cosy lighting that links to cultural norms and identity among Danes. In the weekend or on particular occasions, candles are lit even by people who are well aware of the substantial particle pollution they offer. In other situations, which demand less intensive atmosphere, but still a touch of *hyggelys*, people may instead light an artificial LED candle. Particle pollution, in other words, may both be negatively viewed as dirt and positively viewed as a multi-sensuous co-producer of cosy atmospheres, even by the same individual. The smell of the soot, the slightly heavy air, and the aesthetically un-sterile expression of the warm glow from the subdued light, flickering flames and shadows that orchestrate the intimacy of the cosy atmosphere, shapes a space that does not look sterile, commercialized, or depersonalized, but rather gives a cosy, intimate, and personal touch to a home or space.

Despite my informants' diverse ways of inhabiting the apartments and houses, and amounts of light used, a cosy home is not 'perfect'. It is not something you buy from a shop, but something that is made, is personal, and just a bit *not* too clean. Thus, what is dirty in a bacterial register of knowledge is needed in an atmospheric register of experience. The home needs a bit of what would be considered 'matter out of place' in a bacteriological sense, to avoid the sterile, and thereby not being cosy.

It is in this distinction between cultural perceptions of dirt and the sterile that the *ecstatic* qualities of light, in Böhme's (1995) sense, that the CFL bulb offers, is hitting a nerve. The ecstasy of the bulbs makes things look 'lifeless' and 'sterile' to the informants. Their use of the term 'sterile' may be somewhat metaphorical – more denoting an aesthetic expression, rather than any measurement of bacteria – but nonetheless, it is a minute difference in the light setting of the energy-saving light bulbs that challenges the cosy atmosphere. Shaping aesthetic atmospheres thus wins over environmental ethics. While it may be argued that the CFL bulb is a moralizing technology in a regime of *climate-thinking*, the sterile look it offers is detrimental to the morality embedded in the sensuous regime of *home-making*. The energy-saving light is forcing people to adopt an emerging device paradigm of 'good lighting', which is at odds with the focal thing and practices approach that emphasizes the unsterile, sensuous aspects of lighting, in particular practices of lighting candlelight and fireplaces, although these emit high particle pollution.

As a side effect of ending the 'watt-thinking' to battle energy consumption, a sensory awareness is rising in the wake of climate technologies. It has become increasingly clear how technologies are tincturing the *felt space*; the atmosphere, as that immediate experience, which guides our relationship with the environment and each other. Light has here been normalized as an environmental presence – that is, the way the world appears, and *ought to* appear, that stirs emotions and shapes social life. Such presence is entangled in political, ethical, and cultural spheres of evaluating social actors and places.

Final reflections

In this book, I have explored lighting as atmospheric element and practice in social life. I have shown the homes, values and communities light helps shape and the practices people perform. Central to these arguments has been how atmospheric practices may be the starting point of an analysis of socio-material life where lighting acquires an important position. Such atmospheric practices in the home are formed on the cultural logics and terms that may highlight cosy, bright or excessive light, but also encompass widespread discontent or even dismissal of technologies that would otherwise fit the needs for economic saving and reducing energy consumption.

In this final chapter I have talked about sensory politics of lighting and energy use. On the one hand, it is an empirical task to describe the transition to energy-saving light bulbs, and it is thus important to note, of course, that consumer behaviour and technologies are rapidly changing. On the other hand, the most important aim has been to offer the reader something to think *with*. That point has been to see light as entangled in practices of attuning atmospheres for a variety of other practices to take place, such as social gatherings, cooking, working, relaxing, entertaining, cleaning, or feeling secure. It may be that light offers life in a biological sense, but it appears that we live in the half-light aided by artificial lighting. This points towards the pressing need for an alternative perspective to complement the technical, biological, or psychological approach to lighting that, despite offering crucial knowledge on aspects of light, leaves us little in terms of understanding what counts as good domestic lighting to people, what they do and think about light, and why. Orchestrating light for reading is not simply about the body's ability to see the page, but also about the creating an atmosphere that facilitates a distinction between efficiency-reading, cosy-reading, work-reading, skimming etc. Here candlelight, or lighting a distant corner or adjacent room, are more relevant luminous landscapes to understand than cognitive reading speed, quality or fatigue, if energy consumption is the priority. This shifts the object of study from the *body* to the *spatially felt*, which incorporates the affectively embedded body *in* space and the practices performed *through* space. In this sense, light is not shed *on* the home but *for* the home.

It is now more than ten years ago that Sørensen and I (2007) took the position that the productive role and cultural and social logics of luminosity were under-examined in contemporary society. Light, lighting, darkness, shadows and glare needs to be acknowledged for its active role in making the world sensible to people and help shape their actions and values. The task was to materialize light as analytical lens of understanding the everyday life and the broader issues of space and atmosphere. In a chapter that, in part, focuses on the politics of lighting, it is perhaps also fitting to open up and ask where we are, or should be, going in terms of a fuller understanding of the sociality of light. As discussed in Chapter 2, there has been an upsurge in studies of lighting where people such as Ebbensgaard (2015, 2017), Edensor (2012, 2017; Edensor & Bille 2017; Edensor & Millington 2012), Shaw (2014, 2015), Ingold (2016), Petty (2012; 2010), Pink (Pink & Mackley 2016), Slater's Configuring Light network, and many others, have pushed the agenda much further both in terms of understanding users and producers, and domestic and public spaces. The somewhat culturalistic question about why people across the globe light the way they do, still finds merit to me, since the majority of studies are conducted in Western societies. In this narrower field of everyday life, issues about gender, routines, living together, and the role of lighting technologies in domestic and public spaces, are potentially adding to our limited knowledge about the sociality of light. Furthermore, the productive role of broadening the scope from the object of light and lamp to one of affective powers and atmospheres has recently shed light on the ontology of elements beyond the individual objects, which also calls for a multi-sensuous approach, rather than narrow focus on vision. Norms of lighting – including darkness, shadows, glare, glow and reflections – have historically been different, are different elsewhere, and will be different in the future. I find that answers to how, why and with what social consequences this is the case, are still in much need.

However, I also see the increasing importance for social science to offer answers, concepts and questions in terms of the role of light in shaping and representing class, gender, ethnicity and politics, particularly as it pertains to the development of urban spaces. It is about understanding both the subtle ways lighting norms shift, the commercial influence on urban atmospheres, and the political push to engineer light to solve social issues, beyond gentrification and anti-social behaviour as many studies have shown. But it is also about capturing the fascinating period we are now in, where new lighting technologies are continuously forming our ways of sensing spaces, but where central actors seem largely uncritical towards the social power in such new technologies. There is a task in studying the lighting designers and industry: the way facts and presumptions are made, translated and materialized. That is, get into the machinery of the rapid development of a technology that have defined human history, and now is under fundamental change. This includes developing methods and methodology to better understand the subtleties of meanings and practices that people shape through light.

Finally, I also hope for an increased attention and recognition of the potency of social science and humanities to qualify the role of technology in people's lives, and offer frames or concepts for thinking about lighting – not simply because this may offer answers a lighting designer, for example, may need, but because it offers to complement a dominant focus on the technical, psychological, or more aesthetic approaches, which essentially narrow our understanding of what the complex phenomenon light and lighting is to human life and perception. In such dominant approaches light is cast on an object – the body, a surface, a space – and the behavioural or biological reaction is the point of reference. The point I have made is that people live *with* light, cast it for the environment to make sense, and not just live *in* it.

In the call for understanding the social life of light also lies the political potential. The energy-saving light bulb is doing politics through technologies that have atmospheric consequences. The very same technologies that shape people's experiences, values and behaviour in the world also help curb the climatic changes to the earth's atmosphere. Peter Sloterdijk recognizes the importance of this centrality of technology and atmospheres in politics, when he states that:

> For present-day cultures the question of survival has become a question of the way in which they are reproduced as atmospheric communities. Even physical atmospheres have passed to the stage of their technical producibility. The future era will be climate-technical, and as such technologically oriented. It will be increasingly seen that societies are artificial from the ground up. The air that, together and separately, we breathe can no longer be presupposed. Everything must be produced technically, and the metaphorical atmosphere as much as the physical atmosphere. *Politics will become a department of climate techniques.*
>
> 2011: 245

If this is indeed the case, it highlights how atmospheric practices – the everyday attunement of bodily felt spaces – that include all senses, take centre stage in understanding energy consumption and, by extension, politics. Between the intimate, personal experience of atmospheres – whatever form they may take of *hygge*, serene, intense, uplifting, or heavy – and the distant atmospheric impact of CO_2 emission and particle pollution, the future is indeed atmospheric.

Notes

1 Introduction

1 The term ‘informant’ is used in this book, but without implying that people are passive objects of anthropological study. The term ‘participant’ could also have been used.

2 The sociality of lighting

1 http://energiwiki.dk/index.php/Belysning (accessed 7.4.2016).

2 http://efkm.dk/aktuelt/nyheder/nyheder-2016/juni/ny-undersoegelse-led-lys-vinder-hastigt-frem/ (accessed 29.11.2017).

3 www.standard.co.uk/news/london/the-citys-tallest-ever-tower-under-threat-in-row-over-daylight-a3199266.html (accessed 7.4.2016).

4 www.citymetric.com/horizons/darkness-has-become-luxury-good-london-social-meaning-street-lighting-1504 (accessed 21.3.2016).

3 In the vagueness of *hygge*

1 Unrecorded quote.

2 www.danmarkskanon.dk/om-danmarkskanonen/english/ (accessed 20.11.2017).

3 http://jyllands-posten.dk/debat/breve/ECE9325367/hygge-kan-forsvare-sig-mod-virkeligheden-naar-den-bliver-for-paatraengende/ (accessed 29.1.2017). See also www.theguardian.com/lifeandstyle/2016/nov/22/hygge-conspiracy-denmark-cosiness-trend (accessed 24.5.2018) for more examples.

4 www.danmarkskanon.dk/vaerdi/hygge/ (accessed 20.11.2017).

5 Denmark – The Warm Light of Winter at www.visitdenmark.com/ (accessed 13.8.2008, now removed).

6 www.bolius.dk/derfor-elsker-danskerne-stearinlys-24773/ (accessed 4.4.2016, although that year excluded Sweden, which normally uses around 4 kilos per capita, and Norway, which consumes more but is not part of the EU).

7 http://fdb.dk/analyse/vi-elsker-stearinlys (accessed 27.5.2013).

8 *maGASinet*, January 2015, p. 19.

4 Atmospheric realities

1 www.dictionaryofobscuresorrows.com/post/27720773573/kenopsia (accessed 22.11.2017).

2 *Stimmung* is often translated as 'mood', although such translation is notoriously difficult and words such as state of mind, affectedness and even atmosphere have been used. We also see how the French 'ambiance' in Baudrillard's writing is translated into 'atmosphere' in English (Baudrillard 1968, 1996), while the very word 'ambience' has been through a semantic transformation, from a property of the surroundings – the ambient – to almost bestowing some kind of 'soul' on to a space – ambience (Schmidt 2010; Spitzer 1942a, 1942b). One may either appreciate the conceptual vagueness and plurality, or try to position them in a spectrum where the 'ambient' is a quality of the built environment, *Stimmung* is the individual's way of being in the world merging subject and object, 'aura' emanates from a specified object, charisma from a person, and 'atmosphere' as a focus on spatiality (see Bille et al. 2015; Griffero 2014 for more discussions of differences).

3 As a side note in the terminology of attunement and atmosphere, there is the distinction between the air metaphor of atmospheres – with Böhme associated with the haze, smell and mist (*nebelhaft*, Hauch, Dunst 1995: 22, 27) – and the sound metaphor of attunement – associated with tone, tuning, music. In Danish, '*stemme*' translates into several words associated with sound and balance such as a voice, to tune (e.g. a guitar), tally, vote (both verb and noun), or push back.

5 Atmospheric communities

1 maGASinet, January 2015, p. 19.

2 maGASinet, January 2015, p. 19.

3 Nuances should be mentioned here. During wintertime, many have their curtains drawn since the large window sections offer a thermal bridge, which leaves draughts and cold rooms. Living at the floor level is also a factor where curtains are often drawn, or the lower half of the windows are blurred, to avoid constant exposure to the gazing people walking past. The distance to the neighbour is another issue: the closer the buildings are located, the more often the curtains are drawn in the evening.

6 Qualifying bright lighting

1 4 January 2010.

2 http://denstoredanske.dk/Sprog,_religion_og_filosofi/Religion_og_mystik/Jødedommen/ghetto (accessed 4.11.2016).

3 www.citymetric.com/horizons/darkness-has-become-luxury-good-london-social-meaning-street-lighting-1504 (accessed 28.5.2016).

4 www.urbanplanen.com/#!Mere-lys-over-Remiseparken/c7ba/56bb0d430cf2fd311cdd6ad0 (accessed 28.5.2016, my translation).

5 www.urbanplanen.com/#!Lys-i-m%F8rket-p%E5-fredag/c7ba/01C73041-9985-4F45-9E54-F763310207AD (accessed 28.5.2016, my translation).

6 www.urbanplanen.com/#!1001-lys-t%E6ndes-i-Remiseparken/c7ba/554c5ecb0cf21fee1375c3ef (accessed 28.5.2016, my translation).

7 www.urbanplanen.com/#!Lysfest-og-tryghedsm%F8de/c7ba/5614eaa30cf2a7bb74cc0f47 or www.urbanplanen.com/#!Tilbageblik-p%E5-LYSFEST-15/c7ba/5654d0ae0cf2b6a6e9316775 (accessed 28.5.2016).

7 Ignorance and illumination

1 http://news.cision.com/dk/go--energi/r/for-hamstrede-hun-glodepaerer--nu-er-hun-vild-med-sparepaerer-og-led-paerer,c9226262 (accessed 22.11.2017).

2 This transformation is what Martin Heidegger distinguishes between a ready-to-hand and present-at-hand relationship, or in some cases as un-ready-to-hand, when for instance the lamp switch is broken (Conspicuous), or there is no bulb (Obtrusive), or something is in the way of reaching the lamp (Obstinate) (Heidegger 1996: §16,68).

3 For instance, in Hundested in Northern Sealand, a house is decorated with 20,000 lights, including a large sign: 'Støt Mødrehjælpen' (support the mother helpers), and offering links to online donation webpages (Berlingske. 24.11.2011. 5. Sektion, p. 32).

8 Reflections of sensory politics

1 www.spiegel.de/international/germany/getting-around-the-eu-ban-germans-hoarding-traditional-light-bulbs-a-638494.html (accessed 7.4.2016).

2 www.washingtonpost.com/blogs/innovations/post/energy-innovation-and-the-battle-of-the-bulb/2011/07/14/gIQARsXMEI_blog.html?utm_term=.c6a1d04ce3d8 (accessed 25.11.2017).

3 http://efkm.dk/media/7564/faktaark_led.pdf (accessed 29.11.2017).

4 http://politiken.dk/forbrugogliv/boligogdesign/energi/art5571569/Vi-køber-100.000-forbudte-glødepærer-om-året (accessed 28.11.2017).

Bibliography

Adey, P. 2014. Security atmospheres or the crystallisation of worlds. *Environment and Planning D: Society and Space* 32.
Adey, P. & B. Anderson 2012. Anticipating emergencies: Technologies of preparedness and the matter of security. *Security Dialogue* 43, 99–117.
Al-Salameen, Z. & H. Falahat 2009. Religious practices and beliefs in Wadi Mousa between late 19th and early 20th centuries. *Jordan Journal of History and Archaeology* 3, 175–204.
Albertsen, N. 1999. Urbane Atmosfærer. *Sociologi i dag* 4, 5–29.
Albertsen, N. 2012. Gesturing atmospheres. In *Ambiances in Action: Proceedings of the 2nd International Congress on Ambiances.*
Amit, V. & N. Rapport 2002. *The Trouble with Community: Anthropological Reflections on Movement, Identity and Collectivity.* London: Pluto Press.
Anderson, B. 1983. *Imagined Communities.* London: Verso.
Anderson, B. 2009. Affective atmospheres. *Emotion, Space and Society* 2, 77–81.
Anderson, B. 2010. Morale and the affective geographies of the 'war on terror'. *Cultural Geographies* 17, 219–36.
Anderson, B. & P. Adey 2011. Affect and security: Exercising emergency in 'UK civil contingencies'. *Environment and Planning D: Society and Space* 29, 1092–109.
Anderson, B. 2014. *Encountering Affect: Capacities, Apparatuses, Conditions.* Farnham: Ashgate.
Antonakaki, T. 2007. Lighting and spatial structure in religious architecture: A comparative study of a Byzantine church and an early Ottoman mosque in the city of Thessaloniki. *Proceedings, 6th International Space Syntax Symposium, İstanbul* 0–57, 1–14.
Appadurai, A. 1986. *The Social Life of Things: Commodities in Cultural Perspective.* Cambridge: Cambridge University Press.
Ardener, E. 1992. Ritual og socialt rum. *Tidsskriftet Antropologi* 25, 23–8.
Aries, M. B. C. & G. R. Newsham 2008. Effect of daylight saving time on lighting energy use: A literature review. *Energy Policy* 36, 1858–66.
Armstrong, R. P. 1971. *The Affecting Presence: An Essay in Humanistic Anthropology.* Chicago: University of Illinois Press.
Ash, J. 2013. Rethinking affective atmospheres: Technology, perturbation and space times of the non-human. *Geoforum* 49, 20–8.
Bachelard, G. 1988. *The Flame of a Candle.* Dallas: The Dallas Institute Publications.
Bachelard, G. 1994. *The Poetics of Space: The Classical Look at How We Experience Intimate Places.* Boston: Beacon Press.
Barbaras, R. 1998. *Le tournant de l'expérience.* Paris: Vrin.
Barbaras, R. 2004. Affectivity and movement: The sense of sensing in Erwin Straus. *Phenomenology and the Cognitive Sciences* 3, 215–28.

Barnaby, A. 2016. *Light Touches. Cultural Practices of Illumination, 1800–1900.* London: Routledge.
Bartlett, S. 1993. Shedding light on residential consumers. *Energy* 18, 171–83.
Baudrillard, J. 1968. *Le système des objets.* Paris: Gallimard.
Baudrillard, J. 1996. *The System of Objects.* London: Verso.
Baxendall, M. 1995. *Shadows and Enlightenment.* New Haven: Yale University Press.
Begemann, S. H. A., G. J. van den Beld & A. D. Tenner 1997. Daylight, artificial light and people in an office environment, overview of visual and biological responses. *International Journal of Industrial Ergonomics* 20, 231–9.
Bell, K., L. McCullough, A. Salmon & J. Bell 2010. 'Every space is claimed': Smokers' experiences of tobacco denormalisation. *Sociology of Health and Illness* 32, 914–29.
Bellacasa, M. P. de la 2011. Matters of care in technoscience: Assembling neglected things. *Social Studies of Science* 41, 85–106.
Bellacasa, M. P. de la 2012. 'Nothing comes without its world': Thinking with care. *Sociological Review* 60, 197–216.
Bellacasa, M. P. de la 2017. *Matters of Care: Speculative Ethics in More than Human Worlds.* Minnesota: University of Minnesota Press.
Bijker, W. 1992. *The Social Construction of Fluorescent Lighting or How an Artefact Was Invented in Its Diffusion Stage* (eds) W. Bijker & J. Law. Cambridge: MIT Press.
Bijker, W. 1995. *Of Bicycles, Bakelites, and Bulbs: Towards a Theory of Sociotechnical Change.* Cambridge, MA: MIT Press.
Bijker, W. E., T. P. Hughes & T. Pinch 1987. *The Social Construction of Technological Systems: New Directions in the Sociology and History of Technology.* Cambridge, MA: MIT Press.
Bille, M. 2010. Seeking Providence Through Things. The Words of God versus Black Cumin. In *An Anthropology of Absence. Materializations of Transcendence and Loss* (eds) M. Bille, F. Hastrup & T. F. Sørensen, 167–84. New York: Springer Press.
Bille, M. 2013a. Luminous atmospheres: Energy politics, climate technologies, and cosiness in Denmark. *Ambiances. International Journal of Sensory Environment, Architecture and Urban Space.* DOI: 10.4000/ambiances.376.
Bille, M. 2013b. Verdens uafklarethed. *Tidsskriftet Antropologi* 67.
Bille, M. 2013c. Dealing with dead saints. In *Taming Time, Timing Death. Social Technologies and Ritual* (eds) D. R. Christensen & R. Willerslev, 137–55. Farnham: Ashgate.
Bille, M. 2015a. Hazy worlds: Atmospheric ontologies in Denmark. *Anthropological Theory* 15, 257–74.
Bille, M. 2015b. Lighting up cosy atmospheres in Denmark. *Emotion, Space and Society* 15, 56–63.
Bille, M. 2017. Ecstatic things: The power of light in shaping Bedouin homes. *Home Cultures* 14, 25–49.
Bille, M., P. Bjerregaard & T. F. Sørensen 2015. Staging atmospheres: Materiality, culture and the texture of the in-between. *Emotion, Space & Society* 15, 31–8.
Bille, M., F. Hastrup & T. F. Sorensen 2010. *An Anthropology of Absence: Materializations of Transcendence and Loss* (eds) M. Bille, F. Hastrup & T. F. Sørensen. New York: Springer Press.

Bille, M. & Simonsen, K. in press, Atmospheric practices. On affecting and being affected. *Space and Culture.*
Bille, M. & T. F. Sørensen 2007. An anthropology of luminosity: The agency of light. *Journal of Material Culture* 12, 263–84.
Bille, M. & T. F. Sørensen 2016a. Into the fog of architecture. In *Elements of Architecture. Assembling Archaeology, Atmosphere and the Performance of Building Spaces* (eds) M. Bille & T. F. Sørensen. London: Routledge.
Bille, M. & T. F. Sørensen 2016b. *Elements of Architecture: Assembling Archaeology, Atmosphere and the Performance of Building Spaces.* London: Routledge.
Bille, M. & T. F. Sørensen 2018. In visible presence. In *Oxford Handbook of Light in Archaeology* (eds) C. Papadopoulos & H. Moyes. Oxford: Oxford University Press.
Bille, M. & C. Volf 2011. En åben lukkethed. In *Percipio: Bind II* (eds) M. Ahnfeldt-Mollerup & N. Rostrup, 69–77. København: Kunstakademiets Arkitektskolens Forlag.
Bladh, M. 2011. Energy efficient lighting meets real home life. *Energy Efficiency* 4, 235–45.
Bladh, M. & H. Krantz 2008. Towards a bright future? Household use of electric light: A microlevel study. *Energy Policy* 36, 3521–30.
Blumenberg, H. 1993. Light as a metaphor for truth: At the preliminary stage of philosophical concept formation. In *Modernity and the Hegemony of Vision* (ed.) D. M. Levin, 30–62. Berkeley: University of California Press.
Blunt, A. & R. Dowling 2006. *Home.* London/New York: Routledge.
Böhme, G. 1993. Atmosphere as the fundamental concept of a new aesthetics. *Thesis Eleven* 36, 113–26.
Böhme, G. 1995. *Atmosphäre: Essays zur neuen Ästhetik.* Frankfurt am Main: Suhrkamp.
Böhme, G. 1998a. *Anmutungen: Über das atmosphärische.* Ostfildern vor Stuttgart: Arcaden.
Böhme, G. 1998b. Atmosphere as an aesthetic concept. *Daidalos* 68, 112–15.
Böhme, G. 2001a. *Aisthetik: Vorlesungen über Ästhetik als allgemeine Wahrnehmungslehre.* Munich: W. Fink.
Böhme, G. 2001b. *Aisthetik. Vorlesungen über Ästhetik als allgemeine Wahrnemungslehre.* Munich: Wilhelm Fink.
Böhme, G. 2003. The space of bodily presence and space as a medium of representation. In *Transforming Spaces: The Topological Turn in Technology Studies* (eds) M. Hård, A. Lösch & D. Verdicchio. Online publication of the international conference held in Darmstadt, Germany, 22–24 March 2002.
Böhme, G. 2005. Atmosphere as the subject matter of architecture. In *Herzog & de Meuron: Natural History* (eds) J. Herzog & P. Ursprung, 398–406. Montreal: Lars Müller and Canadian Centre for Architecture.
Böhme, G. 2006a. Atmosphere as the subject matter of architecture. In *Herzog & de Meuron: Natural History* (eds) J. Herzog & P. Ursprung, 340–98. London: Lars Müller Publishers.
Böhme, G. 2006b. *Architektur und Atmosphäre.* München: Wilhelm Fink GmbH.
Böhme, G. 2013a. The art of the stage set as a paradigm for an aesthetics of atmospheres. *Ambiances,* 2–8.
Böhme, G. 2013b. *Atmosphäre: Essays zur neuen Ästhetik.* Frankfurt am Main: Suhrkamp.

Böhme, G. 2017. *The Aesthetics of Atmospheres*. London: Routledge.

Bollnow, O. F. 1941. *Das Wesen der Stimmungen*. Frankfurt am Main.: Vittorio Klostermann.

Bollnow, O. F. 1963. *Mensch und Raum*. Stuttgart: Kohlhammer.

Bolt, B. 2000. Shedding light for the matter. *Hypatia* 15, 202–16.

Borch, C. 2012. *The Politics of Crowds: An Alternative History of Sociology*. Cambridge: Cambridge University Press.

Borch, C. 2014. Introduction. Why atmospheres? In *Architectural Atmospheres: On the Experience and Politics of Architecture* (ed) C. Borch. Basel, Switzerland: Birkhäuser.

Borch, C. 2015. The politics of the senses: Crowd formation through sensory manipulation. In *From Multitude to Crowds: Collective Action and the Media*, 1–18. Berlin: Peter Lang.

Borgmann, A. 1984. *Technology and the Character of Contemporary Life: A Philosophical Inquiry*. Chicago: University of Chicago Press.

Borgmann, A. 1992. Moral Significance of Material Culture. *Inquiry: An Interdisciplinary Journal of Philosophy* 35, 291–300.

Borish, S. 1991. *The Land of the Living: The Danish Folk High Schools and Denmark's Non-violent Path to Modernization*. Grass Valley, CA: Blue Dolphin.

Bouman, M. J. 1987. Luxury and control: The urbanity of street lighting in nineteenth-century cities. *Journal of Urban History* 14, 7–37.

Bourdieu, P. 1984. *Distinction: A Social Critique of the Judgement of Taste*. Cambridge, MA: Harvard University Press.

Bowers, B. 1982. *A History of Electric Light & Power*. London: Peter Peregrinus in association with the Science Museum.

Boyce, P. 2014. *Human Factors in Lighting*. Boca Raton, FL: CRC Press.

Boyce, P. 2017. Editorial: The divorce of the art and science of lighting. *Lighting Research & Technology* 49, 671.

Bradley, R. 1989. Darkness and light in the design of Megalith Tombs. *Oxford Journal of Archaeology* 8, 251–9.

Brennan, T. 2004. *The Transmission of Affect*. New York: Cornell University Press.

Bressani, M. 2015. Light into darkness: Gaslight in nineteenth century Paris. In *Cities of Light: Two Centuries of Urban Illumination* (eds) S. Isenstadt, M. Petty & D. Neumann, 28–36. London: Routledge.

Brinkman, S. & S. Kvale 2014. *InterViews: Learning the Craft of Qualitative Research Interviewing*. Thousand Oaks, CA: SAGE.

Brosnan, M. 1998. *Technophobia. The Psychological Impact of Information Technology*. London: Routledge.

Brox, J. 2010. *Brilliant. The Evolution of Artificial Light*. Boston: Houghton Mifflin Harcourt.

Buchli, V. 1999a. *An Archaeology of Socialism*. Oxford: Berg.

Buchli, V. 2013. *An Anthropology of Architecture*. London: Bloomsbury.

Buser, M. 2013. Thinking through nonrepresentational and affective atmospheres in planning theory and practice. *Planning Theory* 13, 227–43.

Campo, J. E. 1991. *The Other Sides of Paradise. Explorations into the Religious Meanings of Domestic Space in Islam*. Columbia: University of South Carolina Press.

Carrier, J. G. & P. G. Luetchford 2012. *Ethical Consumption. Social Value and Economic Practice*. New York/Oxford: Berghahn.

Carter, S. 2007. *Rise and Shine. Sunlight, Technology and Health*. Oxford: Berg.
Castells, M. 1997. *The Power of Identity: The Information Age: Economy, Society, and Culture, Volume II*. Oxford: Blackwell Publishers.
Chappatte, A. 2014. Night life in southern urban Mali: Being a Muslim maquisard in Bougouni. *Journal of the Royal Anthropological Institute* 20, 526–44.
Chua, L. 2009. To know or not to know? Practices of knowledge and ignorance among Bidayuhs in an 'impurely' Christian world. *Journal of the Royal Anthropological Institute* 15, 332–48.
Cieraad, I. 1999. *At Home: Toward an Anthropology of Domestic Space*. New York: Syracuse University Press.
Cohen, A. P. 1985. *The Symbolic Construction of Community*. London: Routledge (available online: www.bibliovault.org/BV.landing.epl?ISBN=9780226036861).
Cook, M. & T. Edensor 2014. Cycling through dark space: Apprehending landscape otherwise. *Mobilities* 1–19.
Corrodi, M. & K. Spechtenhauser 2008. *Illuminating*. Basel: Birkhäuser.
Cowan, R. S. 1985. How the refrigerator got its hum. In *The Social Shaping of Technology: How the Refrigerator Got Its Hum* (eds) D. MacKenzie & J. Wajcman, 202–18. Milton Keynes: Open University Press.
Croissant, J. L. 2014. Agnotology: Ignorance and absence or towards a sociology of things that aren't there. *Social Epistemology* 28, 4–25.
Cross, J. 2013. The 100th object: Solar lighting technology and humanitarian goods. *Journal of Material Culture* 18, 367–87.
Crowley, J. 2001. *The Invention of Comfort*. Baltimore: Johns Hopkins University Press.
Csordas, T. J. 1994. *Embodiment and Experience: The Existential Ground of Culture and Self*. Cambridge: Cambridge University Press.
Custers, P., Y. de Kort, W. Ijsselsteijn & M. de Kruiff 2010. Lighting in retail environments: Atmosphere perception in the real world. *Lighting Research and Technology* 42, 331–43.
Czeisler, C. A. 2013. Perspective: Casting light on sleep deficiency. *Nature* 497.
Daniels, I. 2015. Feeling at home in contemporary Japan: Space, atmosphere and intimacy. *Emotion, Space and Society* 15, 47–55.
Delanty, G. 2009. *Community. Second edition*. London: Routledge.
Dennis, S. 2016. *Smokefree. A Social, Moral and Political Atmosphere*. London: Bloomsbury.
Desjarlais, R. & C. J. Throop 2011. Phenomenological approaches in anthropology. *Annual Review of Anthropology* 40, 87–102.
Dewalt, K. M. & B. R. Dewalt 2011. *Participant Observation: A Guide for Fieldworkers*. New York: Altamira Press.
Dilley, R. 2010. Reflections on knowledge practices and the problem of ignorance. *Journal of the Royal Anthropological Institute* 16, 176–92.
Dillon, M. 2002. *Artificial Sunshine: A Social History of Domestic Lighting*. London: National Trust Books.
Douglas, M. 2002. *Purity and Danger: An Analysis of Concepts of Pollution and Taboo*. London: Routledge. First published 1966.
Duff, C. 2016. Atmospheres of recovery: Assemblages of health. *Environment and Planning A* 48, 58–74.
Dupont, S. & U. Liberg 2008. *Atmosfære – i pædagogisk arbejde*. København: Akademisk Forlag.

Dutson, C. 2010. *Light Volumes Dark Matter. Patterns for Sustainable Lighting.* London: Helen Hamlyn Centre.
Ebbensgaard, C. L. 2015. Illuminights: A sensory study of iluminated urban environments in Copenhagen. *Space and Culture* 18, 112–31.
Ebbensgaard, C. L. 2017. *Rethinking Urban Lighting: Lighting in Everyday Life.* Roskilde University.
Edberg, P. 2016. *The Cozy Life: Rediscover the Joy of the Simple Things Through the Danish Concept of Hygge.* Createspace.
Edensor, T. 2010. Aurora landscapes: Affective atmospheres of light and dark. In *Conversations with Landscape* (eds) K. Benediktsson & K. A. Lund, 227–40. Farnham: Ashgate.
Edensor, T. 2012. Illuminated atmospheres: Anticipating and reproducing the flow of affective experience in Blackpool. *Environment and Planning D: Society and Space* 30, 1103–22.
Edensor, T. 2013a. The gloomy city: Rethinking the relationship between light and dark. *Urban Studies* 52, 422–38.
Edensor, T. 2013b. Reconnecting with darkness: Gloomy landscapes, lightless places. *Social & Cultural Geography* 14, 446–65.
Edensor, T. 2015a. Introduction to geographies of darkness. *Cultural Geographies* 22, 559–65.
Edensor, T. 2015b. Producing atmospheres at the match: Fan cultures, commercialisation and mood management in English football. *Emotion, Space and Society* 15, 82–9.
Edensor, T. 2017. *From Light to Dark: Daylight, Illumination and Gloom.* Minneapolis/London: University of Minnesota Press.
Edensor, T. & M. Bille 2017. 'Always like never before': Learning from the lumitopia of Tivoli Gardens. *Social & Cultural Geography.* doi.org/10.1080/14649365.2017.1404120.
Edensor, T. & E. Falconer 2015. Dans le noir? Eating in the dark: Sensation and conviviality in a lightless place. *Cultural Geographies* 22, 601–18.
Edensor, T. & S. Millington 2009. Illuminations, class identities and the contested landscapes of Christmas. *Sociology* 43, 103–21.
Edensor, T. & S. Millington 2012. Blackpool Illuminations: Revaluing local cultural production, situated creativity and working-class values. *International Journal of Cultural Policy* 6632, 1–17.
Edgerton, R. B. 1967. *The Cloak of Competence: Stigma in the Lives of the Mentally Retarded.* Berkeley: University of California Press.
Ekrich, R. 2006. *At Day's Close: Night in Times Past.* New York: W. W. Norton & Company.
El-Khoury, R. 2006. Polish and deodorize. Paving the city in late eighteenth century France. In *The Smell Culture Reader* (ed.) J. Drobnick, 18–28. New York: Berg.
Elpidorou, A. & L. Freeman 2015. Affectivity in Heidegger I: Moods and emotions in being and time. *Philosophy Compass* 10, 661–71.
Fennell, C. 2011. 'Project heat' and sensory politics in redeveloping Chicago public housing. *Ethnography* 12(1), 40–64.
Ferreira, B. 2002. *Stimmung bei Heidegger.* Netherlands: Springer.
Ferrier, J. F. 1854. *Institutes of Metaphysic. The Theory of Knowing and Being.* Edinburgh/London: William Blackwood.
Firestein, S. 2012. *Ignorance: How it Drives Science.* New York: Oxford University Press.

Flyvbjerg, B. 2006. Five misunderstandings about case-study research. *Qualitative Inquiry* 12, 219–45.
Foucault, M. 1973. *The Birth of the Clinic*. New York: Pantheon Books.
Frondel, M. & S. Lohmann 2011. The European Commission's light bulb decree: Another costly regulation? *Energy Policy* 39, 3177–81.
Gage, J. 1995. *Colour and Culture: Practice and Meaning from Antiquity to Abstraction*. London: Thames & Hudson.
Galinier, J., A. Monod Becquelin, G. Bordin, et al. 2010. Anthropology of the night. *Current Anthropology* 51, 819–47.
Gandy, M. 2017a. Urban atmospheres. *Cultural Geographies* 24, 353–74.
Gandy, M. 2017b. Negative luminescence. *Annals of the American Association of Geographers* 107, 1090–107.
Garnert, J. 1993. *Anden i lampan: etnologiska perspektiv på ljus och mörker*. Stockholm: Carlsson.
Garnert, J. 1994. Seize the day. Ethnological perspectives on light and darkness. *Ethnologia Scandinavica* 24, 38–59.
Garnert, J. 1997. Über die Kulturgeschichte der Beleutung und des Dunkels. *Historische Anthropologie* 5, 62–82.
Garnert, J. 2011. On the cultural history of Nordic light and lighting. In *Nordic Light: Interpretations in Architecture* (eds) N. Sørensen & P. Haug, 5–13. Denmark: Clausen Grafisk.
Garnert, J. 2016. *Ut ur Mörkret. Ljusets och belysningens kulturhistoria*. Lund: Historiska Media.
Garvey, P. 2005. Domestic boundaries: Privacy, visibility and the Norwegian window. *Journal of Material Culture* 10, 157–76.
Gell, A. 1998. *Art and Agency: An Anthropological Theory*. Oxford: Oxford University Press.
Gershon, I. & D. S. Raj 2000. Special issue, The symbolic Capital of Ignorance. *Social Analysis* 44.
Govoroff, N. C. 1993. The hunter and his gun in Haute-Provence. In *Technological Choices : Transformation in Material Cultures since the Neolithic* (ed.) P. Lemonnier. London: Routledge.
Gram-Hanssen, K. 2008a. Consuming technologies – developing routines. *Journal of Cleaner Production* 16, 1181–9.
Gram-Hanssen, K. 2008b. Energy in homes: An Historical approach to understanding new routines. In *The Culture of Energy* (ed.) M. Rüdiger, 180–99. Newcastle: Cambridge Scholars Press.
Gram-Hanssen, K. 2010. Residential heat comfort practices: Understanding users. *Building Research & Information* 38, 175–86.
Gram-Hanssen, K. 2011. Understanding change and continuity in residential energy consumption. *Journal of Consumer Culture* 11, 61–78.
Gram-Hanssen, K. 2013. Efficient technologies or user behaviour: Which is the more important when reducing households' energy consumption? *Energy Efficiency* 6, 447–57.
Gram-Hanssen, K. 2015. Hverdag i det topisolerede hus. *Råstof* 11–13.
Gram-Hanssen, K. & C. Bech-Danielsen 2012. Creating a new home. Somali, Iraqi and Turkish immigrants and their homes in Danish social housing. *Journal of Housing and the Built Environment* 27, 89–103.
Grasseni, C. 2004. Skilled vision. An apprenticeship in breeding aesthetics. *Social Anthropology* 12, 41–55.

Gregson, N., A. Metcalfe & L. Crewe 2009. Practices of object maintenance and repair. *Journal of Consumer Culture* 9, 248–72.

Griffero, T. 2014. *Atmospheres: Aesthetics of Emotional Spaces*. Farnham: Ashgate.

Gross, M. & L. McGoey 2015. *Routledge International Handbook of Ignorance Studies* (eds) M. Gross & L. McGoeyeds. New York: Routledge.

Großheim, M., S. Kluck & H. Nörenberg 2014. *Kollektive Lebensgefühle: Zur Phänomenologie von Gemeinschaften*. Rostock: Rostock University.

Gullestad, M. 1984. *Kitchen-Table Society: A Case Study of the Family Life and Friendships of Young Working-Class Mothers in Urban Norway*. Oslo Irvington-on-Hudson, New York: Universitetsforlaget; Distribution in United States and Canada, Columbia University Press.

Gullestad, M. 1992. *The Art of Social Relations: Essays on Culture, Social Action and Everyday Life in Modern Norway*. Oslo, Norway New York: Scandinavian University Press; Oxford University Press distributor.

Gullestad, M. 2001. *Kitchen-Table Society: A Case Study of the Family Life and Friendships of Young Working-Class Mothers in Urban Norway*. Oslo: Universitetsforlaget.

Gumbrecht, H. U. 2004. *Production of Presence. What Meaning Cannot Convey*. Stanford: Stanford University Press.

Hand, M. & E. Shove 2004. Orchestrating concepts: Kitchen dynamics and regime change in good housekeeping and ideal home 1922–2002. *Home Cultures* 1, 235–56.

Hansen, J. F. 1976. The proxemics of Danish daily life. *Studies in the Anthropology of Visual Communication* 3, 52–62.

Hansen, J. F. 1980. *We are a Little Land: Cultural Assumptions in Danish Everyday Life*. New York: Arno Press.

Harman, G. 2007. *Heidegger Explained. From Phenomenon to Thing Ideas Explained*. Chicago: Open Court Publishing Company.

Hasse, C. 2008. Postphenomenology: Learning cultural perception in science. *Human Studies* 31, 43–61.

Hasse, C. 2015. *Anthropology of Learning. On Nested Frictions in Cultural Ecologies*. New York: Springer.

Hasse, J. 2008. Stadt und Atmosphäre. *Die alte Stadt* 2.

Hasse, J. 2012. *Atmosphären der Stadt. Aufgespürte Räume*. Berlin: Jovis.

Hasse, J. 2014. Atmospheres as expression of medial power. Understanding atmospheres in urban governance and under self-guidance. *Lebenswelt. Aesthetics and Philosophy of Experience* 4.1, 214–22.

Hauge, B. 2013a. Vinduets betydning. En kvalitativ, antropologisk undersøgelse af hvad vinduet gør for folk (Report).

Hauge, B. 2013b. The air from outside: Getting to know the world through air practices. *Journal of Material Culture* 18, 171–87.

Hauge, B. 2015. Lives under the sun: The sensory qualities of daylight in designing the everyday. *Senses and Society* 10, 71–91.

Hauskeller, M. 1995. *Atmosphären erleben*. Berlin: Akademie Verlag.

Healy, S. 2012. Atmospheres of consumption: Shopping as involuntary vulnerability. *Emotion, Space and Society*.

Heidegger, M. 1971. Building dwelling thinking. In *Poetry, Language, Thought* (ed.) A. Hofstadter, 143–61. New York: Harper & Row.

Heidegger, M. 1996. *Being and Time*. New York: State University of New York Press.

Helliwell, C. 1996. Space and sociality in the Dayak longhouse. In *Things as They Are: New Directions in Phenomenological Anthropology* (ed.) M. Jackson, 128–48. Bloomington: Indiana University Press.
Henare, A., M. Holbraad & S. Wastell 2007a. *Thinking Through Things. Theorising Artefacts Ethnographically*. London/New York: Routledge.
Henare, A., M. Holbraad & S. Wastell 2007b. Introduction. Thinking through things. In *Thinking Through Things. Theorising Artefacts Ethnographically* (eds) A. Henare, M. Holbraad & S. Wastell, 1–31. London/New York: Routledge.
Henckmann, W. 2007. Atmosphäre, Stimmung, Gefühl. In *Atmosphäre(n): Interdisziplinäre Annäherungen an einen unscharfen Begriff* (eds) R. Goetz & S. Graupner, 45–84. Munich: Kopäd.
Hicks, D. & M. C. Beaudry 2010. *Oxford Handbook of Material Culture Studies* (eds) D. Hicks & M. C. Beaudry. Oxford: Oxford University Press.
High, C., A. H. Kelly & J. Mair 2012. *Anthropology of Ignorance* (eds) C. High, A. H. Kelly & J. Mair. New York: Palgrave Macmillan.
Hirschkind, C. 2006. *The Ethical Soundscape. Cassette Sermons and Islamic Counterpublics*. New York: Columbia University Press.
Hobart, M. 1993. *An Anthropological Critique of Development: The Growth of Ignorance* (ed.) M. Hobart. New York: Routledge.
Hobson, K. 2006. Bins, bulbs and shower timers: On the techno-ethics of sustainable living. *Ethics, Place and Environment* 9, 335–54.
Holbraad, M. & M. A. Pedersen 2016. *The Ontological Turn: An Anthropological Exposition*. Cambridge: Cambridge University Press.
Hopkinson, R. G. (ed.) 1964. *Hospital Lighting*. London: Heineman.
van der Horst, H. & J. Messing 2006. 'It's not Dutch to close the curtains': Visual struggles on the threshold between public & private in a multi-ethnic Dutch neighborhood. *Home Cultures* 3, 21–37.
Horta, A., H. Wilhite, L. Schmidt & F. Bartiaux 2014. Socio-technical and cultural approaches to energy consumption: An introduction. *Nature and Culture* 9, 115–21.
Howarth, N. A. A. & J. Rosenow 2014. Banning the bulb: Institutional evolution and the phased ban of incandescent lighting in Germany. *Energy Policy* 67, 737–46.
Howel, J. P. & R. Sundberg 2015. Towards and affective geopolitics: Soft power and the Danish notion of 'hygge'. *Environment, Space, Place* 7, 97–119.
Howes, D. 2004. *Empire of the Senses: The Sensual Culture Reader. Sensory Formations*. Oxford: Berg.
Howes, D. 2006. *Scent, Sound and Synaesthesia. Intersensoriality and Material Culture Theory*. (eds) C. Tilley, W. Keane, S. Küchler, M. Rowlands & P. Spyer. London: Sage.
Howes, D. & C. Classen 2014. *Ways of Sensing*. London: Routledge.
Hughes, T. P. 1985. Edison and electric light (eds) D. MacKenzie & J. Wajcman, 39–52. Milton Keynes: Open University Press.
Huiberts, L. M., K. C. H. J. Smolders & Y. A. W. de Kort 2015. Shining light on memory: Effects of bright light on working memory performance. *Behavioural Brain Research* 294, 234–45.
Hutta, J. S. 2009. Geographies of Geborgenheit: Beyond feelings of safety and the fear of crime. Environment and Planning D: Society and Space 27, 251–74.
Højer, L. 2009. Absent powers. Magic and loss in post socialist Mongolia. *Journal of Royal Anthropological Institute* 15, 575–91.

Højer, M. & M. Vacher 2009. Om hjemblinded og husets antropologi. *Tidsskriftet Antropologi.* 59/60, 3–16.

Haas, J. & K. Vogt 2015. Ignorance and investigation. In *Routledge International Handbook of Ignorance Studies* (eds) M. Gross & L. McGoye. London: Routledge.

Ihde, D. 2010. *Heidegger's Technologies: Postphenomenological Perspectives.* New York: Fordham University Press.

Ingold, T. 2000. *The Perception of the Environment: Essays in Livelihood, Dwelling and Skill.* London: Routledge.

Ingold, T. 2005. The eye of the storm: Visual perception and the weather. *Visual Studies* 20, 97–104.

Ingold, T. 2007a. Earth, sky, wind, and weather. *Journal of the Royal Anthropological Institute* 13, S19–38.

Ingold, T. 2007b. Materials against materiality. *Archaeological Dialogues* 14, 1–16.

Ingold, T. 2011. *Being Alive. Essays on Movement, Knowledge and Description.* London: Routledge.

Ingold, T. 2012. The atmosphere. *Chiasmi International* 14, 75–87.

Ingold, T. 2016. Lighting up the atmosphere. In *Elements of Architecture. Assembling Archaeology, Atmosphere and the Performance of Building Spaces* (eds) M. Bille & T. F. Sørensen. London: Routledge.

Isenstadt, S., M. Petty & D. Neumann 2015. *Cities of Light: Two Centuries of Urban Illumination* (eds) S. Isenstadt, M. Petty & D. Neumann. London: Routledge.

Jackson, M. 1996. *Things As They Are: New Directions in Phenomenological Anthropology* (ed.) M. Jackson. Bloomington: Indiana University Press.

Jay, M. 1993. *Downcast Eyes: The Denigration of Vision in Twentieth-Century French Thought.* Berkeley: University of California Press.

Jenkins, R. 2012. *Being Danish. 2nd issue.* Copenhagen: Museum Tuscelanum Press.

Jensen, C. L. 2013. Sharing conventions for energy efficient lighting? *ECEEE Summer Study Proceedings* 2261–71.

Jensen, C. L. 2014. *What is energy efficient light? A socio-technical analysis of lighting in transition.* Aalborg Universitet, unpublished PhD thesis.

Jensen, O. B., M. Sheller & S. Wind 2015. Together and apart: affective ambiences and negotiation in families' everyday life and mobility. *Mobilities* 10, 363–382.

Jiménez, A. C. & A. Estalella 2013. The atmospheric person: Value, experiment, and 'making neighbors' in Madrid's popular assemblies. *HAU: Journal of Ethnographic Theory* 3, 119–39.

Johansen, B. S. 2015. Locating hatred: On the materiality of emotions. *Emotion, Space and Society* 16, 48–55.

Kapstein, M. T. 2004a. *The Presence of Light: Divine Radiance and Religious Experience.* Chicago: University of Chicago Press.

Kapstein, M. T. 2004b. Introduction. In *The Presence of Light: Divine Radiance and Religious Experience* (ed.) M. T. Kapstein, 1–3. Chicago: University of Chicago Press.

Koslofsky, C. 2002. Court culture and street lighting in seventeenth-century Europe. *Journal of Urban History* 28, 743–68.

Koslofsky, C. 2006. *Evening's Empire: A History of Night in Early Modern Europe. Cambridge.* Cambridge: Cambridge University Press.

Kotamraju, N. P. 2011. Playing stupid, caring for users, and putting on a good show: Feminist acts in usability study work. *Interacting with Computers* 23, 439–46.
Kotler, P. 1973. Atmospherics as a marketing tool. *Journal of Retailing* 49, 48–64.
Kranzberg, M. 1986. Technology and history: 'Kranzberg's Laws'. *Technology and Culture* 27, 544–60.
Kristensen, L. S., S. Traberg-borup, E. Petersen & K. Johnsen 2004. *Lyset i skolen.* Hørsholm: Statens Byggeforskningsinstitut.
Kumar, A. 2015. Cultures of lights. *Geoforum* 65, 59–68.
Laganier, V. & J. van der Pol 2011. *Light and Emotions. Exploring Lighting Cultures. Conversations with Lighting Designers.* Basel: Birkhaüser.
Lash, S. 1994. Reflexivity and its doubles: Structure, aesthetics, community. In *Reflexive Modernization: Politics, Tradition and Aesthetics in the Modern Social Order* (eds) U. Bech, A. Giddens & S. Lash, 110–73. Cambridge: Polity Press.
Latour, B. 1992. '"Where are the missing masses? The sociology of a few mundane artifacts".' In *Shaping Technology/Building Society: Studies in Sociotechnical Change*, 225–58. Cambridge, MA: MIT Press.
Latour, B. 2005. *Reassembling the Social.* Oxford: Oxford University Press.
Law, J. 2002. *Aircraft Stories. Decentering the Object in Technoscience.* Durham: Duke University Press.
Law, J. 2004. *After Methods. Mess in Social Science Research.* London: Routledge.
Law, J. 2011. What's wrong with a one-world world? *Presented at the Center for the Humanities, Wesleyan University, Middletown, Connecticut, 19th September, 2011.*
Law, J. & V. Singleton 2005. Object lessons. *Organization* 12, 331–55.
Lawson, B. 2002. (2002), Healing architecture. *The Architectural Review* 211, 72–5.
Lévi-Strauss, C. 1966. *The Savage Mind.* Chicago: University of Chicago Press.
Levisen, C. 2012. *Cultural Semantics and Social Cognition: A Case Study on the Danish Universe of Meaning.* Berlin: De Gruyter Mouton.
Lewis, A. 2015. Daylighting in older people's housing: Barriers to compliance with current UK guidance. *Lighting Research and Technology* 47, 1–17.
Lewis, A. & J. Torrington 2013. Extra-care housing for people with sight loss: Lighting and design. *Lighting Research & Technology* 45, 345–61.
Leys, R. 2011. The turn to affect: A critique. *Critical Inquiry* 37(3), 434–72.
Linnet, J. T. 2011. Money can't buy me hygge. Danish middle-class consumption, egalitarianism and the sanctity of inner space. *Social Analysis* 55, 21–44.
Love, S. 2016. A sense of architecture in the past: Exploring the sensory experience of architecture in archaeology. In *Elements of Architecture. Assembling Archaeology, Atmosphere and the performance of building spaces* (eds) M. Bille & T. F. Sørensen. London: Routledge.
Lyon, S. & M. Moberg 2010. *Fair Trade and Social Justice. Global Ethnographies.* New York: New York University Press.
Lytken, M. 2016. *Lys og lamper i de danske hjem. Sagkundskabens formidling af en formålstjenlig kunstbelysning.* København, unpublished PhD thesis: Det Kongelige Danske Kunstakademis Skoler for Arkitektur, Design og Konservering.
MacKenzie, D. & J. Wajcman 1999. *The Social Shaping of Technology.* Buckingham: Open University Press.
Marres, N. 2012. *Material Participation. Technology, the Environment and Everyday Publics.* New York: Palgrave MacMillan.

Massumi, B. 2002. *Parables for the Virtual: Movement, Affect, Sensation*. London: Duke University Press.
Mathiasen, N. 2015. *Nordisk lys*. København, unpublished PhD thesis.
Matusiak, B. S. 2012. Nordic daylight. In *Nordic Light and Colour* (eds) B. S. Matusiak & K. F. Anter, 25–38. Trondheim: NTNU – The Faculty of Architecture and Fine Art.
Mauss, M. 1973. Techniques of the body. *Economy and Society* 2, 70–88.
McCormack, D. P. 2008. Engineering affective atmospheres on the moving geographies of the 1897 Andree expedition. *Cultural Geographies* 15, 413–30.
McCracken, G. 2005. Homeyness. A cultural account of one constellation of consumer goods and meanings. In *Culture and Consumption II*, 22–48. Bloomington and Indianapolis: Indiana University Press.
McLuhan, A., D. Pawluch, W. Shaffir & J. Haas 2014. The cloak of incompetence: A neglected concept in the sociology of everyday life. *American Sociologist* 45, 361–87.
McQuire, S. 2005. Immaterial architectures: Urban space and electric light. *Space and Culture* 8, 126–40.
Meier, J., U. Hasenöhrl, K. Krause & M. Pottharst 2014. *Urban Lighting, Light Pollution and Society* (eds) J. Meier, U. Hasenöhrl, K. Krause & M. Pottharst. London: Routledge.
Meneley, A. 1996. *Tournaments of Value: Sociability and Hierarchy in a Yemeni Town*. Toronto: University of Toronto Press.
Merleau-Ponty, M. 1964. Eye and mind. In *The Primacy of Perception and Other Essays on Phenomenological Psychology, The Philosophy of Art, History and Politics* (ed.) J. M. Edie, 159–90. Evanston, IL.: North-Western University Press.
Merleau-Ponty, M. 1968. *The Visible and the Invisible*. Evanston, IL: North-Western University Press.
Miller, D. 2001. *Home Possessions*. Oxford: Berg.
Miller, D. 2005. *Materiality* (ed.) D. Miller. Durham, NC: Duke University Press.
Miller, D. 2007. Stone age or plastic age? *Archaeological Dialogues* 14, 23–7.
Mills, E. 1993. Special issue on lighting. *Energy* 18, 2. P. V–224.
Mills, E. & N. Borg 1999. Trends in recommended lighting levels: An international comparison. *Journal of the Illuminating Engineering Society of North America* 28, 155–63.
Moezzi, M. 1998. Social meanings of electric light: A different history of the United States. *Proceedings of the ACEEE Summer Study on Energy Efficiency in Buildings: Profiting from Energy Efficiency. American Council for an Energy-Efficient Economy* 8, 8123–32.
Mol, A. 2002. *The Body Multiple. Ontology in Medical Practice*. Durham, NC; London: Duke University Press.
Morris, N. J. 2011. Night walking: Darkness and sensory perception in a night-time landscape installation. *Cultural Geographies* 18, 315–42.
Multhauf, L. S. 1985. The light of lamp-lanterns: Street lighting in 17th-century Amsterdam. *Technology and Culture* 26, 236–52.
Navaro-Yashin, Y. 2012. *The Make-Believe Space. Affective Geography in a Post War Polity*. Durham, NC/London: Duke University Press.
Nesbitt, C. 2012. Shaping the sacred: Light and the experience of worship in middle Byzantine churches. *Byzantine and Modern Greek Studies* 36, 139–60.
Norberg-Schulz, C. 1980. *Genius loci*. University of Minnesota: Academy Editions.

Norberg-Schulz, C. 1993. *Nattlandene. Om byggekunst i Norden*. Oslo: Gyldendal Norsk Forlag A/S.
Norman, D. 1988. *The Design of Everyday Things*. New York: Basic Books.
Nye, D. 1990. *Electrifying America. Social Meanings of a New Technology*. Cambridge, MA: MIT Press.
Nye, D. 2010. *When the Lights Went Out: A History of Blackouts in America*. Cambridge, MA: MIT Press.
Nye, D. 2018. *American Illuminations. Urban Lighting, 1800–1920*. Cambridge, MA: MIT Press.
Nørretranders, T. & O. Eliasson 2015. *Lys! Om lys i livet og liv i lyset*. Denmark: Forlaget Tor.
Olesen, B. B. 2010. Ethnic object in domestic interiors. Space, atmosphere and the making of home. *Home Cultures* 7, 25–42.
Olesen, B. & J. Thorndahl 2004. *Da Danske Hjem blev elektriske. 1900–2000*. Viborg: Kvindemuseets Forlag.
Olwig, K. R. 2011. All that is landscape is melted into air: The 'aerography' of ethereal space. *Environment and Planning D: Society and Space* 29, 519–32.
Othman, Z., R. Aird & L. Buys 2015. Privacy, modesty, hospitality, and the design of Muslim homes: A literature review. *Frontiers of Architectural Research* 4, 12–23.
Paddison, S. E. 1995. *Bryggen – den glemte bydel*. Copenhagen: Islangs Brygges lokalhistoriske forening og arkiv.
Pallasmaa, J. 2005. *The Eyes of the Skin*. Chichester: John Wiley & Sons.
Pallasmaa, J. 2011. The ambience of Nordic light. In *Nordic Light, Interpretations in Architecture* (eds) N. Sørensen & P. Haug, 23–33. Denmark: Clausen Grafisk.
Pallasmaa, J. 2014a. Space, place and atmosphere: Peripheral perception in existential experience. In *Architectural Atmospheres: On the Experience and Politics of Architecture* (ed) C. Borch. Basel, Switzerland: Birkhaüser.
Pallasmaa, J. 2014b. Space, place and atmosphere. Emotion and peripheral perception in architectural experience. *Lebenswelt. Aesthetics and Philosophy of Experience* 4, 230–35.
Parrott, F. R. 2005. 'It's not forever': The material culture of hope. *Journal of Material Culture* 10, 245–62.
Pennartz, P. J. J. 1986. Atmosphere at home: A qualitative approach. *Journal of Environmental Psychology* 6, 135–53.
Pérez-Gómez, A. 2016. *Attunement: Architectural Meaning After the Crisis of Modern Science*. Cambridge, MA: MIT Press.
Perkowitz, S. 1996. *Empire of Light: A History of Discovery in Science and Art*. New York: Henry Holt & Company.
Peters, R. 1983. Light and public places. *Places* 1, 41–7.
Petty, M. M. 2010. 'Edge of danger': Electric light and the negotiation of public and private domestic space in Philip Johnson's glass and guest houses. *Interiors* 1(3), 197–217.
Petty, M. M. 2012. Curtains and the soft architecture of the American postwar domestic environment. *Home Cultures* 9, 35–56.
Pfeil, G. 2012. Sarax and the city: Almsgiving and anonymous objects in Dakar, Senegal. In *The Anthropology of Ignorance. An Ethnographic Approach* (eds) C. High, A. H. Kelly & J. Mair, 33–54. New York: Palgrave Macmillan.
Philippopoulos-Mihalopoulos, A. 2013. Atmospheres of law: Senses, affects, lawscapes. *Emotion, Space and Society* 7(1), 35–44.

Philippopoulos-Mihalopoulos, A. 2016. Withdrawing from atmosphere: An ontology of air partitioning and affective engineering. *Environment and Planning D: Society and Space* 34, 150–67.

Philipsen, C. S. 2013. *Hjemfølelse*. Denmark: Roskilde University.

Pile, S. 2010. Emotions and affect in recent human geography. *Transactions of the Institute of British Geographers* 35, 5–20.

Pinch, T. 2009. *The Social Construction of Technology (SCOT): The Old, the New, and the Nonhuman* (ed.) P. Vannini, 45–58. New York: Peter Lang Publishing.

Pinch, T. & W. Bijker 1984. The social construction of facts and artifacts: Or how the sociology of science and the sociology of technology might benefit each other. *Social Studies of Science* 14, 399–441.

Pink, S. 2004. *Home Truths: Gender, Domestic Objects and Everyday Life.* Oxford: Berg.

Pink, S. 2008. Re-thinking contemporary activism: From community to emplaced sociality. *Ethnos* 73, 163–88.

Pink, S. 2009. *Doing Sensory Ethnography.* London: SAGE.

Pink, S., K. Leder Mackley & R. Morosanu 2015. Researching in atmospheres: video and the 'feel' of the mundane. *Visual Communication* 14, 351–69.

Pink, S. & K. L. Mackley 2016. Moving, making and atmosphere: Routines of home as sites for mundane improvisation. *Mobilities* 11, 171–87.

Plummer, H. 2012. *Nordic Light. Modern Scandinavian Architecture.* London: Thames & Hudson.

Proctor, R. & L. L. Schiebinger 2008. *Agnotology: The Making and Unmaking of Ignorance.* Stanford, CA: Stanford University Press.

Rancière, J. 2004. *The Politics of Aesthetics.* London: Continuum.

Rasmussen, S. E. 1964. *Experiencing Architecture* (2nd edn). Cambridge, MA: MIT Press.

Rauh, A. 2012. *Die besondere Atmosphäre: Ästhetische Feldforschungen.* Bielefeld: Transcript.

Rauh, A. 2014. On the ethical-aesthetic potentials of special atmospheres. *Lebenswelt. Aesthetics and Philosophy of Experience* 4, 246–63.

Reich, L. S. 1992. Lighting the path to profit: GE's control of the electric lamp industry, 1892–1941. *Business History* 66, 305–34.

Rescher, N. 2009. *Unknowability: An Inquiry into the Limits of Knowledge.* Plymouth, UK: Lexington Books.

Riedel, F. 2015. Music as atmosphere. Lines of becoming in congretional worship. *Lebenswelt. Aesthetics and Philosophy of Experience* 6, 80–111.

Riedel, F. 2019. Atmosphere. In J. Slaby & C. von Scheve (Eds.), *Affective Societies: Key Concepts.* London & New York: Routledge.

Russell, H. 2016. *The Year of Living Danishly. Uncovering the Secrets of the World's Happiest Country.* London: Icon Books.

Russell, L. S. 1968. *A Heritage of Light: Lamps and Lighting in the Early Canadian Home.* Toronto: University of Toronto Press.

Sangild, T. 2016. 'The Art of Hygge' er blevet en global livsstilstrend. Her er udlandets analyse af stearinlyslandet. *Zetland* Juni.

Saunders, N. 2002. The colours of light: Materiality and chromatic cultures of the Americas. In *Colouring the past: the significance of colour in archaeological research.* (ed.) A. and M. Jones G., 209–26. Oxford: Berg.

Saunders, N. J. 2001. A dark light: Reflections on obsidian in Mesoamerica. *World Archeology* 33, 220–36.
Schivelbusch, W. 1987. The policing of street lighting. *Yale French Studies* 72, 61–74.
Schivelbusch, W. 1988. *Disenchanted Night: The Industrialization of Light in the Nineteenth Century*. Berkeley: University of California Press.
Schmidt, L.-H. & J. E. Kristensen 1986. *Lys, Luft og Renlighed. Den moderne social-hygiejnes fødsel*. Viborg: Akademisk Forlag.
Schmidt, U. 2010. Det ambiente – Sansning, medialisering, omgivelse. Aarhus University Publishing.
Schmitz, H. 2014. *Atmosphären*. Verlag Karl Alber.
Schmitz, H., R. O. Müllan & J. Slaby 2011. Emotions outside the box – the new phenomenology of feeling and corporeality. *Phenomenology and the Cognitive Sciences* 10, 241–59.
Schüll, N. D. 2012. *Addiction by Design. Machine Gambling in Las Vegas*. Princeton, NJ: Princeton University Press.
Schwartz, J. M. 1989. *In Defense of Homesickness*. Copenhagen: Akademisk Forlag.
Seremetakis, C. N. 1994. *The Memory of the Senses, Part I: Marks of the Transitory* (ed.) C. N. Seremetakis. Oxford: Westview Press.
Sharma, A. & T. F. Stafford 2000. The Effect of Retail Atmospherics on Customers' Perceptions of Salespeople and Customer Persuasion: An Empirical Investigation. *Journal of Business Research* 49, 183–91.
Shaw, R. 2013. 'Alive after five': Constructing the neoliberal night in Newcastle upon Tyne. *Urban Studies* 52, 456–70.
Shaw, R. 2014. Streetlighting in England and Wales: New technologies and uncertainty in the assemblage of streetlighting infrastructure. *Environment and Planning A* 46.
Shaw, R. 2015. Controlling darkness: Self, dark and the domestic night. *Cultural Geographies* 22, 585–600.
Shaw, R. 2018. *The Nocturnal City*. London: Routledge.
Shove, E. 2003. *Comfort, Cleanliness and Convenience. The Social Organization of Normality*. Oxford: Berg.
Shove, E. 2010. Beyond the ABC: Climate change policy and theories of social change. *Environment and Planning A* 42, 1273–85.
Shove, E. & M. Pantzar 2005. Consumers, producers and practices: Understanding the invention and reinvention of Nordic walking. *Journal of Consumer Culture* 5, 43–64.
Shove, E. & D. Southerton 2000. Defrosting the freezer: From novelty to convenience. *Journal of Material Culture* 5, 301–19.
Shove, E. & G. Walker 2014. What is energy for? Social practice and energy demand. *Theory, Culture & Society* 31(5), 41–58.
Shove, E., M. Pantzar & M. Watson 2012. *The Dynamics of Social Practice. Everyday Life and How It Changes*. London: Sage.
Shove, E., M. Watson, M. Hand & J. Ingram 2007. *The Design of Everyday Life*. Oxford: Berg.
Simonsen, K. 2007. Practice, spatiality and embodied emotions: An outline of a geography of practice. *Human Affairs* 17.
Sloane, M. 2014. Tuning the space: Investigating the making of atmospheres through interior design practices. *Interiors: Design, Architecture and Culture* 5, 297–314.

Sloterdijk, P. 2016. *Foams. Spheres Volume III: Plural Spherology*. Boston, MA: MIT Press.

Smolders, K. C. H. J. & Y. A. W. de Kort 2014. Bright light and mental fatigue: Effects on alertness, vitality, performance and physiological arousal. *Journal of Environmental Psychology* 39, 77–91.

Smolders, K. C. H. J., Y. A. W. de Kort & P. J. M. Cluitmans 2012. A higher illuminance induces alertness even during office hours: Findings on subjective measures, task performance and heart rate measures. *Physiology and Behavior* 107, 7–16.

Sørensen, E. 2009. *Materiality of Learning. Technology and Knowledge in Educational Practice*. Cambridge: Cambridge University Press.

Sørensen, N. & P. N. Haug 2012. *Nordic light. Interpretations in architecture*. København: Dansk Center for Lys.

Sørensen, T. F. 2012. Delusion and disclosure: Human disposal and the aesthetics of vagueness. *Embodied Knowledge: Historical Perspectives on Technology and Belief*.

Sørensen, T. F. 2015. In praise of vagueness: Uncertainty, ambiguity and archaeological methodology. *Journal of Archaeological Method and Theory* 741–63.

Spitzer, L. 1942a. Milieu and ambiance: An essay in historical semantics, Part I. *Philosophy and Phenomenological Research* 3, 1–42.

Spitzer, L. 1942b. Milieu and ambiance: An essay in historical semantics, Part II. *Philosophy and Phenomenological Research* 3, 169–218.

Spradley, J. P. 1980. *Participant Observation*. New York: Holt, Rinehart and Winston.

Steidle, A. & L. Werth 2013. Freedom from constraints: Darkness and dim illumination promote creativity. *Journal of Environmental Psychology* 35, 67–80.

Stender, M. 2006. Om at bo i glashus, en antropologisk analyse af rum, synliggørelse og beboelsespraksisser i københavnske boliger og arbejdspladser med transparent arkitektur, kandidatspeciale. Copenhagen: University of Copenhagen.

Stephens, A. C. 2015. The affective atmospheres of nationalism. *Cultural Geographies* 1–18.

Stewart, K. 2007. *Ordinary Affects*. Durham, NC/London: Duke University Press.

Stewart, K. 2011. Atmospheric attunements. *Environment and Planning D: Society and Space* 29, 445–53.

Stidsen, L. 2013. *Light Atmosphere in Hospital Wards*. Aalborg: PhD thesis. River Publishers.

Stidsen, L., N. Thuesen & P. H. Kirkegaard 2014. Mapping Danish lighting trends. Sociocultural aspects of light analysed in Bo Bedre. *Nordisk Arkitekturforsking* 1, 117–40.

Stidsen, L., H. S. Bjerrum, P. H. Kirkegaard, N. Thuesen & A. M. Fisker 2011. Patients' light preferences in hospital wards: Related to light atmosphere in Danish homes. *Proceedings of First International Conference on Light 2011*.

Stoller, P. 1989. *The Taste of Ethnographic Things: The Senses in Anthropology*. Philadelphia: University of Pennsylvania Press.

Strauss, E. 1966. *Phenomenological Psychology. The Selected Papers of Erwin W. Strauss*. New York: Basic Books.

Tan, Q. H. 2016. Scent-orship and scent-iments on the scent-ual: The relational geographies of smoke/smell between smokers and non-smokers in Singapore. *The Senses and Society* 11, 177–98.

Tanizaki, J. 2001. *In Praise of Shadows*. New York: Vintage Press.
Tellenbach, H. 1968. *Geschmack und Atmosfäre*. Salzburg: Otto Müller Verlag.
Thibaud, J. P. 2002. From situated perception to urban ambiences. *First International Workshop on Architectural and Urban Ambient Environment, 6–8 February, Nantes, France.*
Thibaud, J. P. 2011. Sensory design. The sensory fabric of urban ambiances. *Senses and Society* 6, 203–15.
Thibaud, J. P. 2014. Urban ambiances as common ground? *Lebenswelt. Aesthetics and Philosophy of Experience.* 4, 282–95.
Thibaud, J. P. 2015. The backstage of urban ambiances: When atmospheres pervade everyday experience. *Emotion, Space and Society* 15, 39–46.
Thibaud, J., J.-P. Thibaud & D. Siret 2012. Embedding ambiance in action International Congress on Ambiances, Montreal.
Thornton, P. 2015. The meaning of light: Seeing and being on the battlefield. *Cultural Geographies* 22, 567–83.
Thrift, N. 2007. *Non-Representational Theory: Space/Politics/Affect*. London and New York: Routledge.
Thrift, N. 2009. Understanding the affectice spaces of political performance. In *Emotion, Place and Culture* (eds) M. Smith, J. Davidson & L. Bondi, 79–96. Aldershot: Ashgate.
Thwing, L. L. 1959. Flickering flames: A history of domestic lighting through the ages. London: G. Bell & Sons.
Tilley, C. 2004. *The Materiality of Stone: Explorations in Landscape Phenomenology*. Oxford: Berg.
Tilley, C. 2007. Materiality in materials. *Archaeological Dialogues* 14, 16–20.
Tilley, C., W. Keane, S. Küchler, M. Rowlands & P. Spyer 2006. *Handbook of Material Culture* (eds) C. Tilley, W. Keane, S. Küchler, M. Rowlands & P. Spyer. London: Sage.
Tisseron, S. 1999. *Comment l'esprit vient aux objets*. Paris: Aubier.
Tonello, G. 2008. Seasonal affective disorder: Lighting research and environmental psychology. *Lighting Research and Technology* 40, 103–10.
Turley, L. W. & R. E. Milliman 2000. Atmospheric effects on shopping behavior: A review of the experimental evidence. *Journal of Business Research* 49, 193–211.
Turner, V. 1966. *The Ritual Process. Structure and Anti-Structure*. Ithaca, NY: Cornell University Press.
Vannini, P. & J. Taggart 2013. Domestic lighting and the off-grid quest for visual comfort. *Environment and Planning D: Society and Space* 31, 1076–90.
Vartanian, O., G. Navarrete, A. Chatterjee, et al. 2013. Impact of contour on aesthetic judgments and approach-avoidance decisions in architecture. *Proceedings of the National Academy of Sciences of the United States of America* 110 Suppl, 10446–53.
Vera, H. 1989. On Dutch windows. *Qualitative Sociology* 12, 215–34.
Verbeek, P.-P. 2011. *Moralizing Technology: Understanding and Designing the Morality of Things*. Chicago: University Of Chicago Press.
Vitebsky, P. 1993. Is death the same everywhere? Contexts of knowing and doubting. In *An Anthropological Critique of Development: The Growth of Ignorance* (ed) M. Hobart, 100–15. London: Routledge.
Viveiros de Castro, E. 1998. Cosmological deixis and Amerindian perspectivism. *Journal of the Royal Anthropological Institute, n.s.* 4, 469–88.

Volf, C. 2013. *Light, Architecture and Health. A method.* Aarhus: PhD thesis, Aarhus School of Architecture.
Wågø, S., B. Hauge & E. Støa 2017. Between indoor and outdoor: Norwegian perceptions of well-being in energy-efficient housing. *Journal of Architectural and Planning Research* 33, 326–46.
Welsh, B. & D. Farrington 2002. Effects of improved street lighting on crime: A systematic review. *Home Office Research Study 251.*
Wetherell, M. 2012. *Affect and Emotion. A New Social Science Understanding.* London: Sage.
Wiessner, P. W. 2014. Embers of society: Firelight talk among the Ju/'hoansi Bushmen. *Proceedings of the National Academy of Sciences* 111, 14027–35 (available online: www.pnas.org/content/111/39/14027.abstract).
Wilhite, H., H. Nakagami, T. Masuda, Y. Yamaga & H. Haneda 1996. A cross-cultural analysis of household energy-use behavior in Japan and Norway. *Energy Policy* 24, 795–803.
Wiil, P. 2016. Den rette stemning – stoflige virkninger i Landsforeningen Bedre Byggeskiks arkitektur. Roskilde University.
Williams, R. 2008. Night spaces: Darkness, deterritorialization, and social control. *Space and Culture* 11, 514–32.
Winther, I. W. 2005. Hjemmet er mere end et hus. *Jordens Folk* 40, 4–9.
Winther, I. W. 2006. *Hjemlighed. Kulturfænomenologiske studier.* Copenhagen: Danmarks Pædagogiske Universitets Forlag.
Winther, T. 2008. *The Impact of Electricity: Development, Desires and Dilemmas.* New York; Oxford: Berghahn Books.
Woodward, I. 2007. *Understanding Material Culture.* London: Sage.
Wu, X. 2008. Men purchase, women use: Coping with domestic electrical appliances in rural China. *East Asian Science, Technology and Society: An international journal* 2, 211–34.
Wu, X. 2009. *Ein Jahrhundert Licht: eine technikethnologische Studie zur Beleuchtung im chinesischen ländlichen Alltag.* Wiesbaden: Otto Harrassowitz Verlag.
Young, D. 2006. *The Colours of Things* (eds) C. Tilley, W. Keane, S. Küchler, M. Rowlands & P. Spyer. London: Sage.
Zahavi, D. 2003. *Fænomenologi.* Frederiksberg: Roskilde Universitetscenter.
Zajonc, A. 1993. *Catching the Light. The Entwined History of Light and Mind.* Oxford: Oxford University Press.
Zardini, M. 2005. *Sense of the City. An Alternate Approach to Urbanism.* Montreal: Lars Müller Publishers.
Zerdy, J. & M. Schweitzer 2016. Provocative atmospheres, refracted nations, and the performance of light. *Performance Matters* 2, 4–22.
Zumthor, P. 2006. *Atmospheres.* Berlin: Birkhäuser Verlag.

Index